W9-AXX-578

The Earth Path

ALSO BY STARHAWK

The Spiral Dance: A Rebirth of the Ancient Religion of the Great Goddess

Dreaming the Dark: Magic, Sex and Politics

Truth or Dare: Encounters with Power, Authority and Mystery

The Fifth Sacred Thing

Walking to Mercury

The Pagan Book of Living and Dying,
with M. Macha Nightmare and the Reclaiming Collective

Circle Round: Raising Children in Goddess Tradition,
with Diane Baker and Anne Hill

The Twelve Wild Swans,
with Hilary Valentine

Webs of Power: Notes from the Global Uprising

The Earth Path

Grounding Your Spirit in the Rhythms of Nature

STARHAWK

HarperSanFrancisco
A Division of HarperCollins*Publishers*

 For information address HarperCollins Publishers, Inc., 10 East 53rd Street, New York, NY 10022.

HarperCollins books may be purchased for educational, business, or sales promotional use. For information please write: Special Markets Department, HarperCollins Publishers, Inc., 10 East 53rd Street, New York, NY 10022.

HarperCollins Web site: http://www.harpercollins.com

HarperCollins®, ®, and HarperSanFrancisco™ are trademarks of HarperCollins Publishers, Inc.

FIRST EDITION

Book design and charts by Kris Tobiassen
Illustrations by Lydia Hess

Library of Congress Cataloging-in-Publication Data is available on request.

ISBN 0-06-000092-9 (cloth)

04 05 06 07 08 RRD(H) 10 9 8 7 6 5 4 3 2 1

Contents

List of Exercises, Meditations, and Rituals

THE SACRED

OBSERVATION

THE CIRCLE OF LIFE

AIR

FIRE

WATER

EARTH

THE CENTER

HEALING THE EARTH

Acknowledgments

This book was inspired and informed by many people. First, let me acknowledge and offer gratitude to the Pomo people, the original people of the land I live on, and to all the indigenous peoples of the earth who have been caretakers, guardians, and keepers of the earth's wisdom for millennia, and whose lands and cultures are continually under siege today.

Susan Davidson introduced me to permaculture many years ago. Penny Livingston-Stark was my first permaculture teacher, is my current teaching partner, and is a dear friend and inspiration. I am grateful also to other teachers, Blythe Reis, Patricia Michael, and Keith Johnson, and to the amazing group at Occidental Arts and Ecology Center: Brock Dolman, Dave Henson, and Adam Wolpert. Erik Ohlsen has grown from a student to a close friend and teaching partner. Abby Wing has been a coteacher and partner in actions.

Bill Mollison and David Holmgren originated the principles and practices of permaculture. Matthew Fox and Brian Swimme showed me the power of uniting science and story. Jon Young and the other teachers of the Wilderness Awareness School opened my eyes and ears and taught me to hear the language of the birds.

Evergreen Erb, Kitty Engleman, and Sunray have cotaught Earth Path workshops with me and helped develop many of the exercises and insights in this book. The many teachers I have worked with in Reclaiming over the years have cocreated and inspired much of this work.

Carol Christ and David Seaborg kindly read chapters and offered helpful feedback and clarifications. Brian Tokar attempted to awaken me to the dangers of biotech many years ago. Luke Anderson helped deepen my understanding and is a tireless activist on the issue, as is Brian.

I am grateful to the committed organizers of the Sacramento Circus, the Greenbloc, and Earth First! and to the forest defenders and all those who have put their lives on the line for the earth. And to all those who have created the many alternatives and positive solutions outlined here.

My partners in our training collective RANT (Root Activist Network of Trainers), Lisa Fithian, Hilary McQuie, Charles Williams, and Ruby Perry, have been a great support throughout the writing of this book. My friends and neighbors in the Cazadero Hills have accompanied me on walks, taught me to recognize plants and animals, shown me mushrooms, and joined me in efforts to protect our lands. Mary Dedanan and Akasha Madron have helped me hold all the complicated logistical threads of my life together, and Ken Genetti is the spider who weaves my Web page. My agent, Ken Sherman, is always a strong support and my editor, Eric Brandt, has been an understanding, flexible, and helpful reader who has sharpened the book's focus. My housemates at Black Cat and my partner, David Miller, have been loving companions on this journey. The children in my life, Kore, Aidan, Allison, Florence, Aminatou, Tijiane, Bowen, Lyra, Johanna, Casey, Emma Lee, Leif, Tashi Sophia, and Ruby, will inherit our efforts at earth-healing and the legacies of our wounds, and this book is dedicated to them.

STARHAWK
CAZADERO HILLS
JANUARY 6, 2004

ONE

Toward the Isle of Birds

On a hilltop in the coastal mountains of northern California, I meet with my neighbors just before sunset on a hot day in July to go to a fire protection ritual. All summer long, our land and homes are at risk for wildfire. In the winter, we get eighty to a hundred inches of rain in a good year, and trees and grasses and shrubs grow tall. But no rain falls from June through September, and in summer the land gets dry as tinder. A small spark from a mower, a carelessly tossed cigarette, a glass bottle full of water that acts as a magnifying lens can all be the beginning of an inferno that could claim our homes and lives.

We live with the constant risk of fire, and also with the knowledge that our land needs fire, craves fire. This land is a fire ecology. All the trees on it evolved in association with forest fires. The redwoods, with their thick, spongy bark, withstand fire. The madrones and bay laurels and tanoaks resprout from root crowns to survive fire. Fire once kept the meadows open, providing habitat for deer and their predators, coyote and cougar. Fire kept the underbrush down, favoring the big trees and reducing disease. The Pomo, the first people of this land, burned it regularly to keep it healthy. As a result, the forest floor was kept open, the fuel load was reduced, and fires were low and relatively cool. But now the woods are dense with shrubby regrowth, the grasses tall and dry. A fire today would not be cool and restorative, but a major inferno.

Below us is the small firehouse that belongs to our Volunteer Fire Department. We can look around to the far horizons and see our at-risk landscape. Deep canyons are filled with redwoods and Douglas firs, with bay laurel and madrone and vast stands of tanoak filling in the open spaces left where stands of giant conifers were logged a hundred years ago and, again, fifty years ago. The tanoaks are bushy, with multiple small stems that create a huge fire hazard. Big-leaf maples line the stream banks, and black oaks stud the open hillsides where fifty years ago sheep grazed. Tall stands of grasses in the open meadows are already dry and ready to burn. Once the meadows would have stayed green all summer with deep-rooted native bunchgrasses, but a century of grazing favored invasive European grasses that wither quickly in the summer heat. Small homes fill the wrinkles in the landscape, most built twenty years ago by back-to-the-landers out of local wood and scrounged materials. On the high ridges, we can see evidence of the latest change in land use, a proliferation of vineyards. Behind us is a huge fallen tree—a remnant of the 1978 wildfire that started just over the ridge and burned thousands of acres.

We begin by sharing some food, talking and laughing together, waiting for everyone to arrive. Then we ground, breathing deeply and with great gratitude the clean air that blows fresh from the ocean just a few ridges over. We imagine our roots going into the earth, feeling the jumble of rock formations and the volatile, shifting ground here just two ridges over from the San Andreas fault. We feel the fire of the liquid lava below our feet, and the sun's fire burning hot above our heads.

We cast our circle by describing the boundaries of the land we wish to protect—from the small town of Cazadero in the east to the *rancheria* of the Kashaya Pomo in the north; from the ocean in the west to the ridges and gulches to the south of us. We invoke the air—the actual breeze we can feel on our skin; the fire, so integral to this landscape yet so dangerous to us now; the water, the vast ocean now covered in a blanket of fog, the sweet springs that feed the land; the earth herself, these jumbled ridges and tall forests.

In the center of the circle is a small bowl. One by one, we bring water from our springs and pour it into the vessel. My neighbors know exactly where their water comes from. Each of us has spent many hours digging out springs, laying water pipes, fixing leaks.

"This is from a spring beyond that hill that flows into Camper Creek that flows into Carson Creek that flows into MacKenzie Creek that flows into Sproul Creek that flows into the South Fork of the Gualala River . . . "

We offer the combined waters to the earth with a prayer of gratitude—great gratitude that we live in one of the few places left on earth where we can drink springwater straight from the ground.

Alexandra has made our fire charm—a circle of bay laurel branches with a triangle lashed within. The triangle is the symbol of fire; the circle represents containment and also the cycle that we know someday needs to be restored. One by one, we come forward and tie on branches we have each brought from trees on our lands. Redwood, from a giant that has withstood many fires. Tanoak, suffering now from a fungal disease that fire might have cured. Madrone, of the beautiful peeling red bark, and buckeye in flower. They are as familiar as our human friends. We know them intimately, know when and how they flower and seed, have watched many individuals grow from seedlings. Some of my neighbors planted these hills after the 1978 fire, worked the creek beds to slow erosion, thinned and released the woods time after time. They know the boundaries of the soil types and the history of each patch of the woods. Ken and Alexandra bring small, uprooted firs, pulled out from a patch on their land where they grow far too thickly for any to get enough light to grow healthy and strong. Once fire would have thinned them—now people do. We add herbs and flowers from our gardens.

We pass the charm around, drumming and chanting to charge it:

> Sacred fire that shapes this land,
> Summer teacher, winter friend,
> Protect us as we learn anew
> To work, to heal, to live with you.
> Green, green crown
> Roots underground.
> Kissed by fire,
> Still growing higher.

Laughing, we dance with the charm, pass it over each other's heads and bodies. These trees and branches are part of us as we have each become part of this land. The water we have brought is our drinking water, the water that grows our gardens. We literally eat and drink the land.

When the charm has gone around, we all hold it together and chant, raising a wordless cone of power, a prayer of protection, and also a prayer for knowledge. We pray that our homes and lives can be preserved as we struggle to learn, once again, how to integrate fire with this land, how to restore the balance that has been so lost.

Then the two young girls who are with us climb the fallen tree behind us and hang the charm high on its branches, where it will overlook the land for the summer. We will see it every time we look up at Firehouse Hill. And when winter comes, and the rain returns, we will take it down and cut it apart in our rain return ritual, where we thank the rain for coming back and pray for the health of the land and the trees. We'll each take pieces of this charm to burn in our woodstoves for our winter fires, to protect our homes in the season when fire warms our hearths and cooks our food.

In thirty or more years of practicing earth-based spirituality, I've probably done thousands of rituals. Some are old and some are new; some have become traditions and some draw on ancient roots. Our fire ritual and rain return ritual are relatively young—we created them less than ten years ago. They don't correspond to the equinoxes or the major Celtic feasts or the indigenous Pomo ceremonies of this land. Yet in some ways they represent the most ancient tradition of ritual and ceremony there is: they are the rituals the land told us to do.

The fire ritual represents, for me, a shift in the way I view my own spirituality. For more than three decades, I've been a Witch, a priestess of the Goddess of birth, growth, death, and regeneration, someone who sees the sacred embodied in the natural world. I've written books, created rituals, and practiced and taught magic, "the art of changing consciousness at will."[1] I've marched, demonstrated, organized, and even gotten arrested trying to protect the integrity of the natural world. Nature has been the heart of my spirituality.

But I grew up a city girl. I didn't spend my childhood roaming the woods and splashing in pristine streams; I spent it playing handball in the parking garage of our apartment in the San Fernando Valley of L.A. There was one good climbing tree in our neighborhood, but it stood in the front yard of a woman who yelled at us to get out every time we got up into its branches. My widowed mother never took us camping, and the summer camps I went to stressed studying Hebrew and saying prayers rather than learning woodcraft. My formal education focused on art and psychology and somehow missed biology and ecology. In something like seven years of higher education, only one course, a class in botany for art majors, taught me anything about observing or interacting with the natural world.

When I began studying, teaching, and writing about Witchcraft and Goddess religion thirty years ago or more, what seemed most important to me is that Wicca (the archaic name for our tradition) valued women, the body, and the erotic. I saw magic as an ancient tradition of psychology, the understanding and training of the human mind. And those are indeed very important aspects of our tradition.

But, as I've celebrated in Pagan communities and lived in both the city and the country, as I've worked in environmental movements and other movements for social and ecological justice, I've come to feel that one aspect of our nature-based religion that too often gets neglected is our actual relationship with nature. To be a Witch, to practice magic, we can't simply honor nature's cycles *in the abstract*. We need to know them intimately and understand them in the physical as well as the psychic world. A real relationship with nature is vital for our magical and spiritual development, and our psychic and spiritual health. It is also a vital base for any work we do to heal the earth and transform the social and political systems that are assaulting her daily.

One of the most rewarding aspects of my own journey over the past decades has been a gradual process of deepening my aesthetic appreciation of nature into real knowledge and true understanding. That process became a journey that was to transform my life, my spirituality, and my understanding of the Goddess. It began my true education, and my transformation from a tourist in nature to an inhabitant—someone who not only loves trees but can plant them, prune them, and understand the complex role they play in everything from soil ecology to weather patterns. Like most eco-activists, I fully confess to being a long-term tree-hugger, and like most Witches, I've always talked to trees. But now, when they talk back, I can assess whether what I'm hearing is truly their message or my own fantasies. I've always loved birds, but now when I hear them call in the treetops around my house, I can often identify their voices and at least guess the general subject of their conversation, even if I can't translate all the details.

This journey also transformed my understanding of the Goddess. For me, now, the Goddess is the name we put on the great processes of birth, growth, death, and regeneration that underlie the living world. The Goddess is the presence of consciousness in all living beings; the Goddess is the great creative force that spun the universe out of coiled strings of probability and set the stars spinning and dancing in spirals that our entwining DNA echoes as it coils, uncoils, and evolves. The names and faces we give the Goddess, the particular aspects she takes, arise originally from the qualities of different places, different climates and ecosystems and economies. In Eleusis, once the most fertile plain in Greece, she was Demeter, Goddess of grain. Up the way, in dry, hilly Athens, she was Athena, Goddess of olives. In Hawaii, she is Pele, Goddess of the volcano. In India, each tribal village has a patron Goddess/devi of its own.

And the tradition we call Wicca arose from people who were indigenous to their own lands. In England, even until recent times, certain families passed on the tradition of "earth-walking," of knowing their own area intimately, understanding the mythological and practical significance of every hill and stream and valley, knowing the uses of the herbs and the medicinal properties of the trees and shrubs, and being responsible for the area's spiritual and ecological health.

David Clarke, in his book *Twilight of the Celtic Gods*, records the story of an informant he calls "the Guardian," who recalls his upbringing in an ancient, earth-based tradition of Yorkshire:

> I come from an old tradition, a very old tradition if the learning passed down from families is to be believed. . . . I was always told that my family and its various branches and offshoots have been in this part of the world since time began . . . and we have worked on the land as farmers, craftsmen and in related professions. . . . Yes, I suppose we are "pagans"—but only in the sense that the world of paganism originally meant the beliefs and practices of those in the countryside. . . .
>
> At the time of my "awakening," as we called it, my maternal grandmother was responsible for passing on the teachings . . . and this at first took the form of what might be called "nature walks"—remember, I was only seven at the time—in which we would walk for miles in all weathers, at all times of year and at all times of day and night. If I tried to speak or ask questions, I was hushed with a "just look and listen" or something similar. . . .
>
> My grandmother explained to me . . . that the earth was a living, breathing entity and everything was interrelated. . . . I had to learn all, and I mean all, the names—local names that is—for every single plant, tree, type of stone, animal, bird, insect, fish and so on. I had to know where they could all be found, what they looked like at any given time of year and what, if any, their uses were—practical, medical or whatever.
>
> . . . I was also eased into the fundamental belief of our tradition—that the land is sacred. And to that end we thought of ourselves as stewards, guardians of the areas where members of our family dwelt, people who could be of some use to others who had forgotten or never knew what we still held on to. . . . Farmers, stockmen, gamekeepers and many ordinary countryfolk all knew of our knowledge of plants and animals, and certain members of the family would help them with natural and herbal remedies for both animal and human problems alike. . . .
>
> The powers that we held in awe were locked inside the landscape, inherent in the power of the weather and manifest in the cycle of the changing of the seasons, and in the end they in turn ran through us.[2]

Our magical practices arose from people who were deeply connected to the natural world, and our rituals were designed to give back to that world, to help maintain its balance along with our human balance. If we leave the natural world out of our practice and rituals in any real sense, if we invoke an abstract earth but never have any real dirt under our fingernails, our spiritual, psychic, and physical health becomes devitalized and deeply unbalanced.

In one sense, this understanding of the Goddess is not new for me. More than two decades ago, I wrote about the Goddess in *The Spiral Dance:* "In the Craft, we do not *believe in* the Goddess, we connect with Her, through the moon, the stars, the ocean, the earth, through trees, animals, through other human beings, through ourselves. She is here. She is the full circle: earth, air, fire, water and essence—body, mind, spirit, emotions, change."[3]

But I understand more deeply now that what we call Goddess or God was the face and voice that people gave to the way the land spoke to them. The rituals and ceremonies and myths of the ancestors all arose from their actual relationship to a specific place on earth. And the tools of magic, that discipline of identifying and shifting consciousness, were the skills of listening to what ethnobotanist Kat Harrison calls "the great conversation,"[4] the ongoing constant communication that surrounds us.

Most of us who live in cities, who are educated to read, write, do arithmetic, and use computers, live our lives surrounded by that conversation yet are unaware of it. We may love nature, we may even profess to worship her, but most of us have barely a clue as to what she is murmuring in the night.

To be a Witch (a practitioner of the Old Religion of the Goddess) or a Pagan (someone who practices an earth-based spiritual tradition) is more than adopting a new set of terms and customs and a wardrobe of flowing gowns. It is to enter a different universe, a world that is alive and dynamic, where everything is part of an interconnected whole, where everything is always speaking to us, if only we have ears to listen. A Witch must not only be familiar with the mystic planes of existence beyond the physical realm; she should also be familiar with the trees and plants and birds and animals of her own backyard, be able to name them, know their uses and habits and what part each plays in the whole. She should understand not just the symbolic aspects of the moon's cycle, but the real functioning of the earth's water and mineral and energy cycles. She should know the importance of ritual in building human community, but also understand the function of mycorrhizal fungi and soil microorganisms in the natural community in which human community is embedded.

In fact, *everybody* should. Our culture is afflicted with a vast disconnection, an abyss of ignorance that becomes apparent whenever an issue involving the natural world arises. As a society, we are daily making decisions and setting policies that have enormous repercussions on the natural world. And those policies are being set by officials and approved by a public who are functionally eco-illiterate.

I was once giving a talk at a university about the need for earth-based spirituality, when I was stopped by a student with a question that stunned me.

"Tell me," the young man asked, "why is the earth important?"

I almost didn't know what to say. I bit back a snide retort—"What planet do you live on?"—and realized with horror that he was quite serious, that somehow all his years of higher education and graduate school had not taught him that we are utterly dependent on the earth for our lives.

"Soil bacteria—they're small things; who cares about them?" said a radio interviewer recently when I was trying to explain why we were protesting a USDA conference promoting genetic engineering to agricultural ministers of the third world. It soon became evident that neither he nor most of the audience understood the difference between genetically modifying an organism and simply breeding plants. If you, the reader, don't yet know that difference or understand why anyone who eats should care about the microorganisms in the soil, by the end of this book, you will.

To develop a real relationship with nature, we don't need to live in the country. In fact, this book and work are very much directed toward city dwellers. The vast majority of us, including the vast majority of Pagans, live in cities. It is in the cities that decisions are made that impact the health and life and balance of the natural world. If you love nature but don't really know her, if you live in the city and find yourself stunned and bewildered in the countryside, or if you perhaps know a lot intellectually about ecology but have trouble integrating your knowledge with your deepest sense of joy and connection, this book can be a guide.

Studying the language of nature can be a dangerous undertaking. For to become literate in nature's idiom, we must challenge our ordinary perceptions and change our consciousness. We must, to some extent, withdraw from many of the underlying assumptions and preoccupations of our culture.

The first set of assumptions are those about the earth and our role in it as humans. One view sees human beings as separate from and above nature. Nature exists as a resource bank that we are entitled to exploit for our own ends. She is of value only in how she can be used for our increased comfort, gain, or profit. This philosophy is held by many religions, but also by both capitalists and classical Marxists. It has resulted in unprecedented destruction of ecosystems and life-support systems all over the planet, from the clearcutting of ancient forests to the building of unsafe nuclear reactors.

But there is a counterpoint to this view, one often held by environmentalists and even some Pagans, that is more subtly destructive. That's the view that human beings are somehow worse than nature, that we are a blight on the planet and she'd be better off without us. In *Webs of Power,* I wrote about this view:

> Now, I admit that a case can be made for this view—nevertheless I think that in its own way it is just as damaging as the worldview of the active despoilers. For if we believe that we are in essence bad for nature, we are profoundly separated from the natural world. We are also subtly relieved of responsibility for listening to the great conversation, for learning to observe and interact and play an active role in nature's healing.

The humans-as-blight vision also is self-defeating in organizing around environmental issues. It's hard to get people enthused about a movement that even unconsciously envisions their extinction as a good. As long as we see humans as separate from nature, whether we place ourselves above or below, we will inevitably create false dichotomies and set up human/nature oppositions in which everyone loses.[5]

A corrective view might arise from the understanding that we are not *separate* from nature but in fact *are* nature. Penny Livingston-Stark, my teaching partner in Earth Activist Trainings that combine permaculture design training with work in earth-based spirituality and activism, often tells the story of her own evolution from believing that we must work *with* nature, to seeing us as working *within* nature, to understanding that we are nature working.

Indigenous cultures have always seen themselves as part of nature. Mabel McKay, Cache Creek Pomo healer, elder, and basketmaker, used to say, "When people don't use the plants, they get scarce. You must use them so they will come up again. All plants are like that. If they're not gathered from, or talked to and cared about, they'll die."[6]

Range management expert Allan Savory describes the vast herds of buffalo and prides of lions that stalked the land he managed in the 1950s in what is now Zambia and Zimbabwe, and he talks about how people coexisted with those creatures:

> People had lived in those areas since time immemorial in clusters of huts away from the main rivers because of the mosquitoes and wet season flooding. Near their huts they kept gardens that they protected from elephants and other raiders by beating drums throughout much of the night. . . . [T]he people hunted and trapped animals throughout the year as well.

Nevertheless, the herds remained strong and the river banks lush and well-covered with vegetation, until the government removed the people in order to make national parks.

> We replaced drum beating, gun firing, gardening and farming people with ecologists, naturalists, and tourists, under strict control to ensure that they did not disturb the animals or the vegetation. . . . Within a few decades miles of riverbank in both valleys were devoid of reeds, fig thickets and most other vegetation. With nothing but the change in behavior in one species these areas became terribly impoverished and are still deteriorating. . . . [T]he change in human behavior changed the behavior of the animals that had naturally feared them, which in turn led to the damage to soils and vegetation.[7]

The indigenous peoples of California burned the forests and grasslands to maintain a mosaic of open meadows and forest cover that was ideal for game. When they dug brodaias for food, they took the larger bulbs and scattered the smaller ones, spreading the stands and giving the young bulbs room to grow. By digging and pruning sedge roots for basketweaving, they encouraged the growth of the sedges that helped protect the soils of the riverbanks. California was a lush landscape, described by early European explorers as abundant with game, wildflowers, birds, fish, and natural beauty. Although the explorers thought they had discovered a pristine wilderness, in reality they had found a landscape so elegantly managed that they were utterly unaware of the human role in maintaining such abundance.

Some indigenous cultures have also hunted animals to extinction and turned fertile land to deserts. I don't want to romanticize other cultures, but I do think it is important to learn from them. On this continent, fire, prayer, ceremony, and myth were all ways indigenous peoples attempted to influence and understand their environment. In a world in which everything a person ate, touched, or used came from the land, humans indeed were part of the land in a deep integration we can only imagine.

Another set of assumptions we must challenge are assumptions about what constitutes knowledge. For centuries, since the start of the "scientific revolution," Western culture has pursued knowledge by breaking a subject or an object into its component parts and studying those parts. We go to doctors who specialize in one organ or one set of diseases (such as cancer or heart disease) or one technique for curing (surgery, psychiatry). We study in universities where we learn biology or chemistry or physics. We've developed a mechanistic, cause-and-effect model of the universe. Compartmentalization has taught us a lot, and produced many advances, but it is only one way of looking at the world. It doesn't allow us to look at the whole, or at the complex web of relationships and patterns that make up a whole.

Science itself has moved beyond the mechanistic model of the universe. Today science is likely to describe the world in terms of networks and probabilities and complexities, as interlocking processes and relationships. Yet our thinking and understanding as a culture does not often reflect this greater sophistication. Nor do our regulations, technologies, and practices.

Magic is, in a sense, pattern-thinking. The world is not a mechanism made up of separate parts, but a whole made up of smaller wholes. In a whole, everything is interconnected and interactive and reflective of the whole—just as in a hologram each separate bit contains an image of the whole. Astrology and Tarot, for example, work because the pattern of the stars at any given moment or the pattern the cards make when they fall reflects the whole of that moment.

Developing a deep relationship with nature means a shift in our thinking, learning to see and understand the whole and its patterns, not just the separate parts.

To really know the Goddess, we must learn to be present in and interact with the natural world that surrounds us, in the city as well as the country or wilderness. Instead of closing our eyes to meditate, we need to open our eyes and observe. Unless our spiritual practice is grounded in a real connection to the natural world, we run the risk of simply manipulating our own internal imagery and missing the real communication taking place all around us. But when we come into our senses, we can know the Goddess not just as symbol but as the physical reality of the living earth.

In developing that real relationship with the Goddess, we also need to reconcile science and spirituality. When our sense of the sacred is based not upon dogma but upon observation and wonder at what is, no contradiction exists between the theories of science and those of faith. As Connie Barlow writes in *Green Space, Green Time*,

> The more we learn about Earth and life processes, the more we are in awe and the deeper the urge to revere the evolutionary forces that give time a direction and the ecological forces that sustain our planetary home. Evolutionary biology delivers an extraordinary gift: a myth of creation and continuity appropriate for our time. . . . Finally, geophysiology, including Gaia theory, has reworked the biosphere into the most ancient and powerful of all living forms—something so much greater than the human that it can evoke a religious response.[8]

When science and spirit are reconciled, the world becomes reenchanted, full of wonder and magic. The great conversation is happening around us in many dimensions. Magic might also be called the art of opening our awareness to the consciousnesses that surround us, the art of conversing in the deep language that nature speaks. And magic teaches us also to break spells, to shatter the ensorcellment that keeps us psychologically locked away from the natural world.

To open up to the outer world, we also undergo inner changes and development. For we are part of the living earth, and to connect with her is to connect with the deepest parts of ourselves. We need the discipline of magic, of consciousness-change, in order to hear and understand what the earth is saying to us. And listening to the earth, doing the rituals the land asks us for, giving back what we are asked for, will also bring us healing, expanded awareness, and intensified life.

Opening up begins with listening. To learn to listen, however, is a long process.

Long ago I read a fairy tale about a prince who learns the language of birds. I remember only the beginning, and though I've searched many times through all my books, I've never been able to find the story again. It begins something like this:

> Once upon a time, there lived a king who had one son, whom he treasured. The king wanted to give his child every advantage, so he sent him to be educated on the sacred Isle of Birds, where he could learn the language of birds. After seven years the prince returned.
>
> "What have you learned?" the king asked.
>
> "I can hear something," the prince replied.
>
> "What! That's all? After seven years?" The king was irate. "You'd better go back and study harder."
>
> So the prince went back to the island, and, after another seven years, he returned home.
>
> "What have you learned?" the king asked again.
>
> "I can hear something, and I can understand something," the prince replied.
>
> "What! That's all? After twice seven years?" Again the king was furious, and sent his son back to the island.
>
> After another seven years, the son returned home again.
>
> "What have you learned?" the king asked, somewhat wearily this time.
>
> "Well, I can hear something, I can understand something, and I can say something," the prince replied.
>
> Angered beyond words, the king threw his son out into the wide world, and the prince was forced to make his way alone.

The story continues, but that opening has much to teach us. To begin with, it implies that true education is about learning the language of nature, and it's a slow process. To learn the language of birds takes time. It took the prince seven years just to *hear* something. We need to slow down, to learn to see and listen, to sharpen our powers of observation. Your admission fee to the Isle of Birds is simply the willingness to set aside some time in your life to be in nature—whether that's an alpine meadow in the wilderness or a vacant lot in the inner city.

Once we have learned to hear, then we can begin to understand. And only after we understand do we begin to speak, to intervene.

The story speaks to a core principle of one of the other disciplines that has deeply informed my relationship with nature: the system of ecological design known as permaculture. Developed by Bill Mollison and David Holmgren, the term *permaculture* comes from both "permanent agriculture" and "permanent culture." It includes principles, practices, and ethics that enable us to design sustainable environments that function like natural systems—for growing food but also for growing human community. Penny Livingston-Stark and Blythe Daniels taught the design course I took in 1996, and since then Penny and I have collaborated on Earth Activist Trainings that combine a permaculture design course with training in earth-based spirituality and the skills of organizing and activism.

Permaculture teaches us to begin with long and careful observation rather than careless intervention. We begin by taking the time to hear and see something, and then look for ways to make the least possible change for the greatest effect. We make small changes first, and observe their effect.

Another discipline that has influenced my ability to hear and understand has been the training I've received from the Wilderness Awareness School, a self-study program directed by Jon Young, which teaches tracking and actually has a course in learning the language of birds. I've been an erratic student, but the routines and approaches I've learned have deeply changed my way of being in the wilderness and the world.

Both those disciplines, along with three decades or more of magical practice and teaching, inform this book. My hope is that it can be a trail that takes you out into nature, that deepens and informs your magical practice as well as your daily life, and that helps ground earth-based spirituality in real ground, real earth.

My intention in this book is to do four things. First, to suggest practices and exercises that will teach us to observe, to hear something. Second, to help us understand something about how the natural world works by engaging our hearts and spirits as well as our minds—through storytelling, myth-making, and trance journeys. Third, to help us learn to speak, to create nature-based ritual, to communicate back to the living beings around us. And finally, to help us act: to know what solutions exist, to understand the practical ways that we can transform our way of living to be more in harmony with the earth, and to have a sound basis for actions in defense of the earth.

I'm writing this book on a computer powered by the sun through the solar panels on my cabin—but I don't expect every reader and every Witch to run out and convert their electrical system tomorrow. But I do expect that every reader will come away with an expanded understanding of the palette of alternative energy options, the implications of our public policy regarding energy use for the continued health of the earth and the general state of conflict in human society—and the vital relevance of these issues to a spirituality based on nature.

I'm eating food from my garden, but I don't expect that every reader and every Witch will grow their own vegetables. I do expect that readers will come away with a deeper appreciation of the Goddess as the complex cycle of birth, growth, decay, and regeneration that makes for soil fertility; with the realization that growing food and eating food are spiritual acts; and with an understanding of how the decisions we make about our food and agricultural systems impact the viability of the earth and human society.

We are animals, evolved to live in a vibrant, thriving, diverse world. It is our birthright to know pristine old-growth forests, wildflower-studded prairies, clear streams, and skies split open by the flight of falcons. The rising of the ocean from global warming is ultimately more real, and more important to the web of life on the planet, than the rising of stock prices or profit margins. The complex exchange of nutrients in the soil is more vital to life on earth than any negotiated trade agreement.

When we learn to hear and begin to understand, then the environment becomes real to us. We can start to speak: to interact sensitively with the natural world, to create visionary solutions to our problems. With all the dire crises and potential disasters that surround us, it lies within our human power to create economies and societies that can provide for our needs sustainably, that can create shared abundance while healing and restoring the environment around us, and that can nurture human freedom and creativity along with natural diversity and health. Ultimately, the test of our education comes in our ability to work with nature to transform our world.

Close this book. Walk outside, if you can, or at least go to a window and open it. Close your eyes and sniff the air. Listen. Who do you hear calling on the wind? Are the birds chattering? Are the tree frogs chanting in chorus? Do you hear the rhythmic throb of city traffic? The cycling trill of car alarms? The cries of children at play?

Everything around us is always speaking. We can heal only by first learning to hear, to understand, and, in time, to respond. As we do, the world becomes richer, a more complex and vibrant place. Open your eyes; see the patterns of light and shadow, the play of the wind. You have already begun your education in the language of nature. You have already set foot upon the Isle of Birds, which is always right here, wherever we are.

TWO

Seeds and Weapons

How We View the World

Early in the morning of June 21, 2003, a phone call awakened those of us staying in the organizers' house for the Sacramento protests against the conference organized by U.S. Secretary of Agriculture Ann Veneman and the USDA to promote biotech and industrial agriculture to ministers from WTO countries, in the run-up to the Cancun ministerial scheduled for September.

"They're raiding the Welcome Center!" a frantic voice told us. "There are a dozen cops and a paddy wagon. Come down!"

Three of us—Lisa, Bernadette, and I—had our clothes on in minutes and were in the car, racing to downtown Sacramento. We arrived at the Welcome Center, a warehouse with a large parking lot next to it, to find masses of police and a huge paddy wagon circling. The police, it turned out, had not actually obtained a search warrant or entered the center. They were entirely occupied with the dangerous materials they found in the parking lot: a bucket of nails and two buckets of seedballs made in the permaculture workshop the day before.

Seedballs are a technique for planting on abandoned and inhospitable ground. You take a variety of seeds, designed to create a "guild" (a self-sustaining mini-community of plants), roll them up in mud containing some compost and a high degree of clay, and then strew the seedballs over the ground you want to plant. The mud and clay protect the seeds from being eaten by birds, and when the rains come, the clay helps hold moisture, enabling the seeds to germinate.

These particular seedballs had been made the day before in a workshop led by Erik Ohlsen[1] and openly attended by the public and the media. They contained a mixture of native wildflowers, legumes (members of the bean and pea family that fix nitrogen and provide fertility), along with mustards and daikon radishes (to build biomass and to put deep roots into the ground and retrieve nutrients that had leached deep below). All the seeds were organic.

Bernadette and I tried to explain this to the officers on the scene, but it was clear to me that we weren't getting through. In part, we faced the same difficulty with the police that we do with the general public around issues of biotech and agriculture: a lack of understanding of the basic principles of ecology. More than that—the whole biotech industry, and the larger system of corporate industrial agriculture that it's part of, is based on a different model of the world than the one that inspired the making of the seedballs.

Industrial agriculture comes out of a mechanistic model. A plant is seen as a product, needing specific inputs of various chemicals, and soil as a stabilizing base to hold it up. Anything in that soil that is not the desired product is seen as competition, to be eliminated. Bugs and pests and diseases should also be attacked and eliminated. It's a worldview of simple causes and effects: if Bug A eats your plant, kill it and your plant will grow. If weeds compete with your corn, kill them (and everything else in the soil) and your plant will grow better. If what you want is corn, plant as much of it as you can, choosing the one variety that will produce the highest yield, so that you can maximize your true crop—profit.

This model extends to the way we view the genetic heritage of the planet. One cause produces one effect: one gene produces one trait. Therefore, why not insert the gene from a flounder, say, into a tomato, to increase its levels of protein? Why not alter soybeans to withstand herbicides so you can plant them and conveniently kill everything else?

The mechanistic model assumes that the world is knowable and controllable. Unintended consequences of an action are seen as anomalies, not "real" consequences, and therefore often go unseen, unacknowledged, and unaccounted for. "Proof" is the drawing of a clear line of simple cause and effect. This has great advantages for corporations bent on making profit. A large corporation can clearcut a hillside and spray herbicides on the exposed ground that get into the water supply: the landslides below, the cancers that arise in the community that lives nearby, the loss of the salmon that once spawned in the stream, go unaccounted for. They are "externalities," unintended consequences. Monsanto can release genetically modified canola that pollutes an organic farmer's fields with its pollen, but Monsanto does not have to add that cost to its accounts. (In fact, Monsanto can sue the farmer for royalties!)

This model is being widely sold to us as "science." It's high-tech, it's postmodern, it's the cutting edge, it will feed the world, and anyone who objects to it is accused of clinging to some romantic past.

But in reality, this model is nineteenth-century science. Science itself began to move beyond it somewhere back in the 1920s, when Heisenberg discovered the uncertainty principle and Einstein began cooking up his theories.

Einstein's theory of relativity showed that matter and energy were one seamless whole, and Heisenberg proved that the observer inevitably affects what she observes. Linear, singular cause and effect was left behind even in the thinking of many nineteenth-century scientists, such as Darwin, whose theory of evolution dealt with complex interrelationships.

The unintended consequences of applying this model to meeting our basic needs are devastating. The "Green Revolution" of the 1970s is a prime example. By applying simplistic science, technology, industrial models, and corporate structure to the agriculture of the third world, we were told, food production would increase and starvation and poverty would end. In reality, the opposite happened. Green Revolution varieties increased yields only when used in conjunction with chemical fertilizers and pesticides that destroyed the health of the soil and the community, while yielding great profits for their manufacturers. Hundreds of local varieties of rice, wheat, and corn were replaced by one or two hybrids, and much biodiversity—the fruits of thousands of years of local selection and adaptation—was lost.

Vandana Shiva, Indian social justice activist and ecofeminist, writes about the "miracle" seeds:

> In the absence of additional inputs of fertilizers and water, the new seeds perform worse than indigenous varieties. The gain in output is insignificant compared to the increase in inputs. The measurement of output is also biased by restricting it to the marketable elements of crops. But, in a country like India, crops have traditionally been bred to produce not just food for humans, but fodder for animals and organic fertilizer for soils. In the breeding strategy for the Green Revolution, multiple uses of plant biomass seem to have been consciously sacrificed for a single use. An increase in the marketable output of grain has been achieved at the cost of a decrease in the biomass available for animals and soils from, for example, stems and leaves, and a decrease in ecosystem productivity due to the over-use of resources.[2]

The Green Revolution is one example of current agricultural practices that favor a "weaponry" approach to agriculture, killing pests with toxic chemicals, tackling weeds with herbicides, and destroying soil life with chemical fertilizers. And these practices don't work: insect damage to crops has increased by twenty percent with the introduction of chemical pesticides since the 1940s.[3] These practices have destroyed farming communities from Iowa to India, driving small farmers off the land and consolidating land and food production in corporate hands.

The model represented by the seedballs comes out of the worldview being articulated by twenty-first-century science. Systems, complexity, chaos, and Gaia theories are some of its manifestations, but it is also much older, akin to the way indigenous peoples have always experienced the earth as alive and relational. "We had so many relatives," said Mihilikawna elder Lucy Smith, "and we all had to live together; so we'd better learn how to get along together. The plants, animals, birds—everything on this earth. They are our relatives and we better know how to act around them or they'll get after us."[4]

This view sees the world as a complex and dynamic web of relationships. There are no simple causes and effects: any change in the web reverberates and affects the whole; small changes can become amplified to have large effects that cannot be predicted. This is sometimes called the "butterfly effect" of chaos theory, from the analogy that a butterfly flapping its wings in Brazil could produce a tornado in Texas.

In this model, a plant is part of a living community of relationships that includes billions of soil microorganisms, worms, insects, other plants, birds, predators, and humans, all of which interact together to create a network of dynamic interactions. A crop can't be seen in isolation—it is part of the web. So our seedballs contained not just one kind of seed, but the nucleus of a group of plants that could coexist in beneficial relationships with each other, which would also benefit the health of the soil and provide conditions for increasing diversity and complexity.

This model looks at systems, not isolated elements. If bugs are devouring your plants, it's a sign that something is out of balance in the overall community. Some predator that could eat the bugs is missing, or something is putting the plants under stress and making them more vulnerable. If your plants are diseased, look to the health of the soil.

In the dynamic web model of the world, we understand that every action or change has a myriad of effects, intended and unintended. The world is not completely knowable or controllable—it's filled with complexities that go beyond our comprehension, with wonder and mystery. And because it is complex, because causes and effects are linked in networks rather than simple lines, the same act will not always produce the same effect. In making changes, therefore, we need to be responsible for any potential reverberations and careful not to produce large-scale damaging and/or irreversible effects. We do this by starting small, by carefully monitoring the changes we produce, and by making the least change necessary to produce a result.

From the dynamic worldview, genetic engineering as currently practiced is a travesty on many counts. First, genetically modifying our food plants risks unintended and irreversible consequences on a staggering, global scale. Already in southern Mexico the wild stands of teosinte, the ancestor of corn, are polluted with bioengineered genes. We simply have no way of knowing what this might mean in the long run. A precious source of biodiversity, of potential change and evolution, has been affected irreversibly.

The wild parent plants of our food plants contain the full genetic potential, the original wild vigor, the unexpressed possibilities inherent in the species. The contamination of the ancestor of corn means that potential is now diminished or lost. It also shows that there is no effective way to quarantine genetically engineered plants that, like corn, pollinate on the wind. When we discover adverse health or environmental effects from a genetic modification, there is no practical way to recall that modification from the environment.

Moreover, the assumption that one gene controls one trait is not borne out by current research. A 2002 press release from the Center for the Biology of Natural Systems (CBNS) at Queen's College, New York, described a review of scientific literature conducted by Dr. Barry Commoner, director of the Critical Genetics Project at CBNS. That review, which was subsequently published in the February 2002 issue of *Harper's* magazine,

> cites a number of recent studies "that have broken the DNA gene's exclusive franchise on the molecular explanation of inheritance." [Commoner] warns that "experimental data, shorn of dogmatic theories, point to the irreducible complexity of the living cell, which suggests that any artificially altered genetic system must sooner or later give rise to unintended, potentially disastrous consequences."
>
> Commoner charges that the central dogma—that one gene equals one trait—a seductively simple explanation of heredity, has led most molecular geneticists to believe it was "too good not to be true." As a result, the central dogma has been immune to the revisions called for by the growing array of contradictory data, allowing the biotechnology industry to unwittingly impose massive, scientifically unsound practices on agriculture.
>
> Commoner's research sounds a public alarm concerning the processes by which agricultural biotechnology companies genetically modify food crops.

> Scientists simply assume the genes they insert into these plants always produce only the desired effect with no other impact on the plant's genetics. However, recent studies show that the plant's own genes can be disrupted in transgenic plants. Such outcomes are undetected because there is little or no governmental regulation of the industry.[5]

In June 2003 Commoner himself said,

> The living cell is not merely a sack of chemicals, but a unique network of interacting components, dynamic yet sufficiently stable to survive. The living cell is made fit to survive by evolution; the marvelously intricate behavior of the nucleoprotein site of DNA synthesis is as much a product of natural selection as the bee and the buttercup. In moving DNA from one species to another, biotechnology has broken into the harmony that evolution produces, within and among species, over many millions of years of experimentation. Genetic modification is a process of very *unnatural* selection, a way to perversely reinvent the inharmonious arrangements that evolution has long ago discarded.[6]

In a worldview of simple cause and effect, we test for "safety" by testing for the effects we can anticipate or predict. But we *can't* test for the safety of effects we haven't anticipated.

In an ominous case, a German biotech company engineered a common soil bacterium, *Klebsiella planticola*, to break down wood and plant wastes and produce ethanol. It passed all its safety tests—until Michael Holmes, a graduate student at Oregon State University, decided to test it in living soil and discovered that all the plants sprouted in that soil died. Worse, it persisted in the soil, as do other genetically modified bacteria. Had it been released for use, it might have spread and, according to geneticist David Suzuki, could conceivably have wiped out all plant life on the continent.[7]

With truly dangerous organisms such as that floating around, it was somewhat surprising to see the level of fear and alarm that our innocent organic seedballs generated in the Sacramento police. They decided, after consultation with their superiors, that we could keep our bucket of nails, since we appeared to be engaged in various building projects rather than producing bombs or planning to hijack airplanes with them. However, they insisted on confiscating the seedballs as "projectile weapons."

It was clear to me that the police basically didn't understand the seedballs, and therefore were afraid of them. They had no category in their minds for "way of planting complex community of beneficial relationships," whereas they did have a category for small, round objects that could be thrown. In fact, they were looking for weapons, eager to find something that could justify the millions of dollars and massive deployment of personnel, the collection of stun guns, tear-gas guns, pepper-spray guns, rubber-bullet guns, M16s, horses, clubs, and armored personnel carriers with which they intended to protect the city from our hordes of puppet-carriers and potentially illegal gardeners.

Looking for weapons, they found our seedballs and perceived them as such. They then spent quite a bit of the day back at the station testing their ballistic capabilities, for the evening news featured cops throwing seedballs at Styrofoam walls and commenting on how they "exploded on contact."

We, on the other hand, had clearly not thought of our seedballs as weapons, or we wouldn't have left them out in plain sight in the parking lot to dry. So in a sense the police action expanded our thinking. In permaculture, we try to get multiple uses for each element in a system. Sometimes that's difficult—a rose, for example, looks pretty and its thorns might discourage intruders from an area, but aside from that most hybrids are not greatly useful in the garden. However, if I think about them as potential weapons, their uses are myriad—the prickly stalks could be used to attack unarmed civilians, the thorns could be inserted into the tires of police cars, the hips lobbed with slingshots at the windows of McDonald's. . . . And think about the lethal potential of something bigger—say, an apple tree!

Ironically, the empty boxes the police had brought to load up the seedballs were marked "Explosives," "Pepper Spray Balls," and "Rubber Bullets." Since they had turned our seeds into weapons, I felt that it would only be fair to do the reverse. But I've tried it and it doesn't work: no matter how many pepper-spray balls you bury, you won't get a single chile pepper, and planting rubber bullets won't produce any rubber trees.

The animate model of the universe is probably the most ancient way of experiencing and being in the world. Yoruba priestess Luisah Teish describes this mode of consciousness: "Prior to the white colonization of the continent, West Africans believed in an animated universe, in the process I call 'Continuous Creation.' Continuous creation means that the generation and recycling of energy is always in effect."[8]

Okanagan artist Jeanette Armstrong says, "We know there's an old, old entity that we are all just minute parts of. We are all just disturbances on the surface of that old entity we could say is humanity. We add to that consciousness continuously."[9]

The man described in Chapter One as "the Guardian," inheritor of an ancient land-based tradition in Britain, says,

> Fundamentally, the belief that was handed down to me was this: that the world and everything in it was driven by an awesome power which could be seen—but only by its effects. This power was generally considered to be female. . . . We didn't need to make representations of her like statues and the like because she was all around, everywhere. . . . [W]hy have statues and such when the whole valley you lived in can be seen as the living body of the mother on which we lived?[10]

While indigenous cultures are all different, one thing they share in common is a perception of the world as alive and themselves as embedded in a matrix of complex relationships. Myth, ritual, ceremony, prayer, and offerings are tools cultures use to maintain a balance between the human and nonhuman communities. In many parts of the world, that view remains intact. All over the world, indigenous cultures are struggling to retain their lands and way of life in the face of an assault by cultures based on very different values. Since that assault originates in what we call Western culture, it's worth looking a bit into our own origins in the West.

Archaeologist Marija Gimbutas, in her many books and excavations, documented that the origins of European civilization, too, lay in cultures that honored the earth and valued cooperation over ruthless competition and war.[11] Their depictions of the sacred expressed in art, pottery, sculpture, and architecture were images of nature and natural cycles, plants, animals, birds, fish, and insects, and of the Goddess—the birth-giving, nurturing, and death-wielding regenerative force of life.

In *Truth or Dare: Encounters with Power, Authority, and Mystery*, I explored some of the long story of how that organic, holistic worldview was replaced by one which took war as its ruling metaphor and divided power by gender.[12] That clash, between matrifocal, woman/earth-centered culture and patriarchal, male-ruled culture, has been going on for over five thousand years in Europe and the Middle East. Much of Western culture can be seen as a dialogue between those strands. "Pagan" cultures often reflected both worldviews: the Celts, for example, were a warlike, chieftain society that retained myths and legends and rites honoring the cycles of nature and awarding a high status to women.

When Christianity came into Europe, it incorporated many of the earlier, nature-centered traditions. The Virgin Mary took the place of the Mother Goddess, churches were built on ancient sacred sites, holy wells were attributed to saints rather than Gods and Goddesses, but the old practices remained. Healing traditions that came from knowledge of the land, the plants, and their properties, and traditions of divination and prophecy, spells and charms and hexes, lingered on as they do today in Latin America, where *curanderas*, traditional spiritual healers, practice alongside the priests of the Catholic Church. Those who passed on and practiced the old ways were called Witches, from an Anglo-Saxon root meaning "to bend or twist." They were the ones who could bend fate and twist the future into favorable paths.

So remnants of indigenous traditions survived in the form of healing traditions—herbalism, which gave us many of our modern medicines, as well as naturopathic and chiropractic medicine. The old religion also remained as folk customs and beliefs, songs, dances, and stories. The fairy tales we today relegate to children were originally stories for adults, the surviving myths and wisdom teachings of earlier cultures.

In the sixteenth and seventeenth centuries in Europe, new economic stresses caused by the influx of gold from the Americas challenged the power of the old ruling classes, which was based on land. A new power began to arise, based on money, trade, and the beginnings of capitalism. With it came a new ideology, the mechanistic model of the universe, which saw the world as made up of separate objects that had no inherent life, could be viewed and examined in isolation from one another, and could be exploited without constraint.

For this new economic order to be accepted, old ideas of the dynamic interrelatedness of the universe and the sacredness of nature needed to be broken down. A new ideology was enforced, and one mechanism for effecting this mass change in consciousness was the fear and terror engendered by the Witch burnings.

The sixteenth and seventeenth centuries were the prime era of Witch persecutions, when first the Catholic and later the Protestant churches attacked all that remained of the old traditions of healing and magic, and the earlier understanding of the world as alive, animate, and speaking.

Today there is much debate about exactly how many Witches were killed, and it is likely that the numbers are far below the nine million that we once postulated. But the impact of the Witch trials was nonetheless enormous.

Anyone could be suspected of being a Witch, and once accused, people found it difficult or impossible to prove the contrary. The Church, both Catholic and Protestant, defined Witchcraft as traffic with the devil, and clerics and their minions tortured suspects. In England, where the use of torture was limited, sleep deprivation, starvation, and rape were employed. On the Continent, human ingenuity was horrifically twisted to invent new ways to deliver pain: the rack, the thumbscrew, the bastinado, and other creative implements of torture were applied. People were fed suggestions of what they had done, and forced to confess. They were tortured until they implicated others, so that no one in a community was safe. The persecutions tended to focus on the peasant and working classes, however, dying away once they reached the upper echelons of society. Two-thirds of the victims were women.

Most of the victims were not actually Witches—that is, were not practitioners of the remnants of the pre-Christian, earth-based religions and healing traditions. Most were simply unfortunates, targeted because of some quarrel with a neighbor or because they perhaps owned a bit of land someone else coveted.

The Witch persecutions did not do away with the old beliefs and practices. Many still survive, even today. In rural France many villages still have a traditional healer, each of whom might specialize in a different malady. For a stiff neck, you might go to one village; for a stomachache, to another. When my friend Rose fell off a ladder in a small village in the Lot region of southern France, we took her to a *roboteuse* who manipulated her neck and cured her. These traditions are passed down in families, alternating genders in each generation.

My friend Ellen Marit was a traditional Sami shaman, from the north of Norway. Her people are called Lapps by outsiders, but they call themselves Sami. She had learned her healing traditions from her father and was passing them on to one of her sons. Their traditions included drumming, trancework, visits to special places of power, and energetic healing. Tragically, her son was murdered while still learning ancient skills, and she herself died a few years later, of a combination of grief and stomach cancer caused by the 1986 nuclear accident in Chernobyl, which strongly affected the reindeer that are the traditional food of the Sami.

Marija Gimbutas came from Lithuania, the land last to be Christianized in western Europe, where ancient Pagan traditions survive to this day, and were strongly alive in the 1920s and 1930s of her childhood. She spoke of seeing peasants kiss the ground each morning, and of how they perceived "Mother Earth as lawgiver. You didn't spit on her or strike her, especially in the spring when she was pregnant," but honored her.[13] Guardian trees were protected, and sacred snakes were fed.

In Ireland, on a walk through the Burren, the bleak but beautiful granite landscape of west County Clare, the women in our tour group visiting sacred sites are impressed with the wide-ranging knowledge of our guide, who has several advanced degrees in botany and biology. At one point, one of the women suggests we do a ritual in an Iron Age ring fort on the top of the hill.

"Not if I'm with you, you don't," our guide says. "That's where the Little People live. You don't mess with the Little People. I'm a farmer, and I need the sun to shine and the rain to fall and my cattle to give milk, and someday I might marry and I need my wife to have children. I'm not messing with the Little People!"

If these beliefs and traditions still persist today, imagine how much stronger they were centuries ago. These traditions maintained an animate and dynamic worldview, and strengthened people's attachment to place and to what was left of communal and tribal attachments to the land.

In the sixteenth and seventeenth centuries, there were still areas of common land in Europe that belonged to the community rather than to individuals. While landownership was highly concentrated and enormously hierarchical, land was nevertheless not considered mere property that could be bought or sold in isolation—but rather a nexus of rights and responsibilities deeply tied to a community. Peasants might not own any land, but they might have the hereditary right to gather wood in the lord's forest or graze their pigs under his oak trees. The folk customs—the maypoles and Morris dances and fairy tales tied to specific places on the landscape—all reinforced those traditions. Again, even today in Ireland old sites are left undisturbed, certain hilltops undeveloped, and road crews detour around certain trees where the Little People are said to live. The view of the land as animated by spirits and nonhuman intelligences was a deterrent to its wholesale exploitation.

The animate worldview and the way of life it represented were targeted by the Witch persecutions, which had several key impacts. First, they broke some of those ties to the land and attacked the underlying worldview by labeling all traffic with and attunement to those other voices as devil worship. They helped pave the way for the enclosure of the commons, the privatization of what had once been collectively held—a process which continues on today through global trade agreements and development. They also undermined the solidarity of the peasant class, which had mounted a series of rebellions over centuries.

Second, they were an attack on forms of knowledge and healing that did not have the approval of the authorities. Midwives, herbalists, and traditional healers, many of whom were women, were considered suspect, and the practice of medicine became a specialized activity concentrated in the hands of male doctors. Although the herbalists of that time were more empirical and truly "scientific" than the doctors of the day (who were busy bleeding people according to their astrological signs), the doctors' knowledge was considered official and valid while the midwives' and herbalists' knowledge was seen as superstition or outright traffic with the devil.

Finally, they were an attack on women. Most of the victims were women, and the evils of the satanic worship that the Church claimed to find were directly attributed to the generally evil nature of women. This justified increased repression of women and restriction of women's roles.

I've written at length about this period in *Dreaming the Dark: Magic, Sex, and Politics*,[14] and don't want to repeat that essay here. But I do want to examine the impact of the Witch persecutions as it still affects us today.

People often ask me, even after I've spent thirty years in this field, why I use words like "Witch" and "magic." I use "Witch" to identify with the heritage outlined above, to place myself firmly in the line of outlaw healers and purveyors of unapproved wisdom. And I use the word "magic" for much the same reason. I could say "sophisticated non-mechanistic psychology," but that term lacks the same ring.

Magic is a discipline of the mind, and it begins with understanding how consciousness is shaped and how our view of reality is constructed. Since the time of the Witch persecutions, knowledge that derives from the worldview of an animate, interconnected, dynamic universe is considered suspect—either outright evil or simply woo-woo.

But whenever an area of knowledge is considered suspect, our minds are constricted. The universe is too big, too complex, too ever-changing for us to know it completely, so we choose to view it through a certain frame—one that screens out pieces of information that conflict with the categories in our minds. The narrower that frame, the more we screen out, the less we are capable of understanding or doing. The police, in the incident that begins this chapter, could not see our seedballs as anything but potential weapons, because that is the frame they were looking through.

Our culture includes some major framemakers. The media is one; academia is another. Aviv Lavi, an Israeli journalist who participated on a panel I was on in Tel Aviv, spoke about how the media determines not just the content of what is said, but the framework of what it is possible to think or talk about. He described how for many years the possibility of a unilateral Israeli withdrawal from Lebanon was not even mentioned. The pros and cons were not debated: it just wasn't a subject for serious consideration—until in May of 2000 Peres decided to do it, and suddenly everyone had been advocating it all along.

Academia does the same. Having worked for ten years on a film about Marija Gimbutas's life, I've become aware of the tremendous backlash against her work among scholars. She is criticized, sometimes by those who haven't actually read her work, for "leaping to conclusions," lacking evidence, interpreting rather than reporting, not following the rules of academic proof, which requires clear lines of cause and effect.

Marija's scholarship reflects the dynamic worldview she wrote about. She looked for patterns, not simple causes and effects, and the encyclopedic breadth of her knowledge allowed her to see wholes, not simply examine the details of an isolated part. Of course, her conclusions can be questioned, and like any human being, she made errors. But the overall whole that she uncovered and revealed is grounded on what she saw and documented of early cultures, and it expands our sense of human possibility.

When we use language that fits into the established framework of the culture, when we try to make our ideas respectable, we limit what we can say and think. But when we use a term like "magic," when we leap out of the constrictions of respectability and cease to care if people see us as woo-woo, suddenly we can think about *anything*. We expand the range of our inquiry beyond the categories already fixed in our minds.

So the Gaia theorists, wanting to be accepted in the realm of science, are always very careful to say that despite the Goddess name for their theory, they are not talking about the earth as a being with consciousness, but about the earth as a self-regulating system that functions "like" an organism. But as soon as we proclaim ourselves Witches and start talking about magic, all the serious scientists turn their backs and leave us free to contemplate Gaia's consciousness and listen for her messages.

U.S. defense minister Donald Rumsfeld, when pressured about the missing weapons of mass destruction that were the Bush administration's main pretext for attacking Iraq, responded in February 2003 by saying, "As we know, there are known knowns. There are things we know we know. We also know there are known unknowns. That is to say we know there are some things we do not know. But there are also unknown unknowns, the ones we don't know we don't know."[15]

He's been laughed at for that statement, but actually I think it's perhaps the most illuminating thing he has ever said publicly. The media and academia deal in known knowns, and are also quite comfortable with known unknowns. But the really interesting questions are the unknown unknowns, and "magic" lets us contemplate that realm.

So we decide, on that rainy hillside in the Burren, to clap our hands three times and say, "I do believe in fairies." And we don't do a ritual inside the ancient ring fort, but look for another spot away from it. Before we begin, we address the spirits of the land and the Little People with an offering, expressing our respect. Immediately the sun comes out for the first time that day, and a raven flies overhead, cawing. The sun remains out through the duration of our ritual, and as soon as we're done, the rain begins again. There's no way to account for this by any current scientific theory I know of, and no way to "prove" that it had anything to do with our decision about the ritual or our address to the spirits. But as we continue to address the land with respect on our journey, and to note that the rain stops each time we begin a ritual, we identify a pattern. Some unknown unknown has entered the picture.

Does magic work? Not by waving a wand, Harry Potter style, and muttering the right incantation for the right result. Not by any simple sense of cause and effect. But magic does work, in the terms of its own worldview. Which is to say, once we understand the universe as a dynamic whole—a whole that we, with our human minds, are part of—we also understand that any change in any aspect of the whole affects the whole. Magic, then, is the art of discerning, choosing, and attuning onself to those changes.

Going back to our ritual on the hillside, we could say that we and our guide and the hills and rocks, the ring fort, the other living communities of that hillside, the raven, the sun and rain and clouds, the history and legends, and the spirits and energies and unknown consciousnesses around us are part of a whole that is a particular moment in the universe. If our guide were not there to warn us about the ring fort, we might be in a different whole. Once our whole includes the consciousness of the Little People, it opens the possibility of many forms of communication. Or we could say that the sunshine and the weather changes and the raven are a reflection of the harmonious whole that our acts of respect helped to form.

Whenever we are able to live for a moment within that consciousness of the whole, we become more whole, more healed. How sad, how grim and tragic, to live within an awareness that can see a seedball only as a weapon, that misses the wonder of the potential for transformation and growth within it, that is forced to view the world as an arena of danger, combat, and betrayal. In today's world we are more and more pressured to live within that limited consciousness, to accept its restrictions as reality, to discount any other possibilities as fantasy, romanticism, wishful thinking. Yet it is the constriction of our imagination that produces that grim world.

Today we live in a world so devitalized, so alienated and fragmented, that many of us are hungry for magic: for a way to perceive and experience the whole, to live in a dynamic, animate universe. It's no wonder that Harry Potter is popular worldwide among adults as well as children, for a world of talking hats and whomping willows and flying broomsticks reflects that sense we have as children that everything around us is alive and has a consciousness of its own. Instead of forcing children to "outgrow" that awareness, we should cherish it as the vital understanding that can help us become healers of this wounded world. For we cannot intervene effectively, cannot say something back to the world, unless we first understand and hear something. And we cannot hear unless we open our ears and realize that the world is speaking to us.

And as soon as we do, we become more alive, more wild, more at home in a vital and dynamic universe.

THREE

The Sacred

Earth-Centered Values

One morning I was sitting on my back deck, meditating on the question of how to make change in the world. The forest was all around me and I was asking the question "Can you change a system from within? Or from without?"

"Systems don't change from within," I heard the forest say. "Systems try to maintain themselves."

I figured that the forest, being a complex system itself, ought to know. But to say, "The forest told me," is already to create a simplified frame. It's a frame I find useful: it's a way of perceiving that's comfortable for my human awareness and allows me to hear something I might otherwise miss. But it is also a simplification of a larger framework, one that might perceive me and my mind and my question and the forest around me and the moment that includes my long-term relationship with that particular spot as a whole in which my mind and the forest's mind are not separate beings talking to each other but one process that together produced that insight.

Magic is itself a framework. Indeed, human beings cannot walk around, function, and continue to tie our shoes without putting some kind of simplified frame around the overwhelming whole of the world. Perhaps only enlightened buddhas can truly remove all frames from the world and exist in ultimate reality.

What follows is my own framework, my understanding of the values and principles that derive from a Goddess-centered view of the world. But, of course, part of the essence of that view is respect for diversity, and for the spiritual authority inherent in each person. Other Witches and Goddess *thea-logians* (*thea* as in "Goddess" instead of *theos* as in "God") may frame values and issues very differently.

Magic teaches us to be aware that we are viewing the world through a frame, warns us not to confuse it with ultimate reality or mistake the map for the territory. Moreover, part of our magical discipline is to make conscious choices about which frame we adopt.

As soon as we start making choices, we have entered the realm of values. The criteria we use for choosing one frame over another come from what we ultimately value most, what we consider sacred. To consider something sacred is to say that it is profoundly important, that it has a value in and of itself that goes beyond our immediate comfort or convenience, that we don't want to see it diminished or denigrated in any way. The word "sacred" comes from the same root as "sacrifice"—because to choose any one value is to relinquish another. If something is sacred to us, we are willing to sacrifice something to protect it, willing to take a stand or to risk ourselves in its service. We don't idealize sacrifice, however. Aligning ourselves with what is truly sacred means serving those things that also feed and renew us, that give us the greatest joy and pleasure, that evoke our deepest love.

As Witches we have a huge responsibility, because we are polytheists. We see many great powers and constellations of energies in the universe that we call Goddesses and Gods, and we choose which we will worship or ally ourselves with. We do also see the underlying unity and oneness of the universe, but being a polytheist is a way of acknowledging that no one name or description or spiritual path can do justice to that immense whole. Gods and Goddesses and sacred texts and religions are all frames, descriptions, maps. No one of them is the whole landscape itself.

But if we have no sacred text, no Ten Commandments, no Ultimate Authority, how do we know what to value? If we don't see the world as a simple battle of good versus evil but as an interplay of forces and counterforces seeking a dynamic equilibrium, on what do we base our ethics?

If we see the world as a dynamic whole, then the first question we might ask when we face a choice is "How does this action or decision impact the whole?" That's not a simple question to answer, because the whole is beyond our complete knowledge, and acts have unexpected consequences.

And how do we know if an effect is beneficial or not? Since earth-based spirituality takes nature as its frame, we can look to natural systems as a model. To understand whether something is beneficial, we need to understand what constitutes health in a natural system, and to know something about how ecosystems work.

A healthy ecosystem might be one that is characterized by cooperative and interdependent relationships among its members, and that is diverse and complex enough to be resilient, to maintain itself in the face of change. Energy and resources are spread throughout the system so that diversity can thrive. No more energy or resources are used to maintain the system than come in from the sun or are generated by the life processes of the system itself. Members of the ecological community are free to express in their unique ways the great creative energies of the universe.

Although we often think of nature as "red in tooth and claw," a field of ruthless competition for survival, today's more sophisticated understanding of ecology sees an enormous amount of cooperation and interdependence. In a forest, trees grow in conjunction with mycorrhizal fungi that interpenetrate the root hairs and extend their ability to take in food and nutrients. Voles and flying squirrels eat the fungi and excrete the spores, spreading them throughout the woods. Through the network of fungi, trees can nurture their own young, and trees in the sun share nutrients with trees in the shade, even those of different species.

In a natural system, the right level of diversity and complexity increases health and resilience. A prairie, which might have hundreds of species of grasses, forbs (or broad-leafed plants), legumes, and flowers in a single square yard, is far more diverse than a field of genetically engineered corn. If a new disease arises, it might affect one or a few of the prairie plants, but hundreds of others would survive. The ground would still be covered: the billions of soil bacteria and the worms below ground would still live. But if a new disease attacks the modified corn plants, they might all die. The exposed soil would erode, with devastating consequences to the below-soil life.

But that healthy diversity lies within a certain spectrum. If we tried to increase it by planting bananas and mangos in an Iowa prairie, obviously those newcomers would die, because they require different growing conditions. Healthy diversity is the maximum diversity that can adapt to the local conditions of life. Those differences in local adaptation create the larger mosaic of biodiversity over the earth-whole.

Abundance, or the provision of resources and energy so that members of an ecological community can thrive, is also a value. Abundance is constrained by sustainability, the need for a system to be self-replenishing, to not consume more than it can create. The margin of abundance is the free gift of the sun's energy, which is constantly showered on the earth, the only true margin of profit that exists. To benefit the whole, that abundance must be spread around and shared, not concentrated so that a few elements have most or all of the resources and others lack what they need.

Freedom and creativity are, perhaps, human values, but they are also aspects of a healthy natural system. A healthy system is dynamic, not static, ever-changing and adapting and evolving. If members of an eco-community are controlled or restricted from expressing their potential or making choices, their ability to adapt is limited. Life has shown, again and again, that it is enormously creative, and alignment with that creativity is one of the marks of health.

There are other human values that we might want to include in our definition of what constitutes "benefit": love, compassion, gratitude, joy—all characteristics that arise in the presence of a healthy, vibrant whole. But love and compassion are more—we might think of them as part of the earth-whole's immune response to dis-ease. They are the emotions that mobilize us as human beings to care for and nurture something, to heal a hurt, to right a wrong.

When a system is whole and healthy, when it is based on relationships of interdependence and cooperation that further resilience, diversity, abundance, sustainability, creativity, and freedom, it exhibits that balance we humans call "justice."

Once we have a model in our minds for what health looks and feels like, we can ask ourselves, when contemplating any act or decision, "Will this create beneficial relationships?"[1]

Answering this question, like the earlier questions, is not as simple as it might seem, because to decide if a relationship is beneficial, especially to the whole, we need to understand something about how systems work. Magic has some guiding principles to offer, and so do systems theory, permaculture, and ecology. What follows is a synthesis of all of these.

The Interplay of Consciousness, Energy, and Form

The universe is a whole, made up of many smaller wholes, circles within circles. How we define those wholes and draw their boundaries profoundly affects how we perceive them and how energy moves within them.

The world is an interplay between consciousness, energy, and matter or form. We know that energy can be transformed into matter and that the atoms of matter can be split to release enormous energy. Matter certainly affects consciousness: try being happy when you don't have enough to eat, or feeling a great sense of well-being while being hit on the head. Consciousness also affects matter: some decision I've made, some image in my mind, is moving me to hit you on the head.

Magic teaches us that consciousness can direct energy in both overt and subtle ways and that energy-flows set the patterns that result in manifestation or form.

Because everything is interdependent, there are no simple, single causes and effects. Every action creates not just an equal and opposite reaction, but a web of reverberating consequences. Everything we do affects the whole.

Every whole is made up of an interplay between consciousness, energy, and form. The universe is infused with consciousness—*is* consciousness, shifting and changing and dancing.

Every consciousness is always communicating. The more we open ourselves to hear and understand that communication, the more we can begin to speak back.

The language of that communication may not be words; it may be emotions, energies, scents, images, events. In speaking back, we also need to move beyond words.

How Energy Moves

Energy moves in cycles, circles, spirals, vortexes, whirls, pulsations, waves, and rhythms—rarely if ever in simple straight lines.

Abundance in a system comes not just from how much energy or resources flow in, but how many times that energy and those resources recirculate before flowing out. If the water you use to wash your dishes is reused to water the garden, you have double the amount of effective water. In an abundant system, waste is food; pollution is an unused resource.

Some of those cycles are *self-constraining*. (In systems theory these are called "negative feedback cycles," but people who associate "negative feedback" with criticism find that term confusing.) Self-constraining or self-regulating cycles work like the temperature regulation system in your body. When you get too hot, you begin to sweat, and the evaporation of the sweat cools you down. When you get too cold, you begin to shiver to warm up. Your body generally does a good job of maintaining an equilibrium, a base temperature. Living systems are characterized by many self-constraining cycles. Gaia as a planetary organism also includes self-regulating cycles.

Other cycles are *self-reinforcing*, or *self-amplifying*. (In systems theory, these are called "positive feedback cycles," although their effects are not always positive.) Self-reinforcing cycles can work like a good composting system: I compost my garden and kitchen wastes, which produces more fertility in the garden, which produces more wastes to compost, and so on. Or they can work like an addiction: I drink too much, so I don't show up at work, so I get fired, so I feel bad about myself, so I drink more, and so on. While self-*constraining* cycles help maintain equilibrium, self-*reinforcing* cycles are driving engines of change, for better or worse. Sometimes self-reinforcing cycles continue until the system reaches a new equilibrium—for example, I reach the absolute limit of how much my garden can produce at a heightened level of fertility. Sometimes, when they have negative effects, they continue until the system crashes, having used up its available resources—for example, I run out of unemployment insurance, friends I can borrow from, couches I can stay on, and I "hit bottom" as an alcoholic.

Energy imbalances in a system create turbulence, movement in spirals and vortexes, which evens out the spread of energy throughout the system.

Form and Matter

Form reflects underlying flows of energy. Nature is full of patterns, or forms, that repeat because they reflect ways that energy flows. Spheres, circles, branches, spirals, waves, and radials are common patterns. Trees, river systems, and the blood in our veins share a branching pattern; snail shells and pea tendrils spiral; water, light, and sand dunes travel in waves. Observing, understanding, and using these patterns can help us direct energy more effectively and create healthier systems. (In Chapter Eleven, we will delve more deeply into the mystery of patterns.)

Form is more rigid, fixed, and resistant to change than energy. The health and function of a system depend not just on what is there, but on where each element is in relation to everything else, and on when each element enters and leaves the system. The right thing in the wrong place, or at the wrong time, can be devastating. When something is in the right place at the right time, it can perform more than one function. A comfrey plant in the midst of your most fertile garden bed will take over and crowd out your vegetables. But on the edge of your garden, it can serve as a barrier to encroaching grass and provide you with medicinal poultices and healing herb tea. And that's not all. It can reclaim lost nutrients from deep soil layers. Its leaves make an excellent mulch or addition to a compost pile, and are good chicken fodder; fermented, they produce a juice that can be diluted and used as a liquid fertilizer or foliar feed for plants. Comfrey flowers feed bees and beneficial insects and are a delicious addition to a salad.

In an abundant system, each element performs multiple functions. We can do more with less, as Buckminster Fuller was fond of saying.

In a secure and stable system, each necessary function is performed by more than one element. If one thing fails, a backup can perform its function.

Making Beneficial Choices

These principles may seem abstract, but throughout this book we will be seeing examples of how they manifest and how we might apply them. But for now, let's go back to our discussion of values and decisions. Knowing what we mean by "beneficial," and coming from some basic understandings about how consciousness, energy, and matter work, how might we apply these to choices we make?

We each make decisions all the time, small ones and large ones. Do I spend an extra dollar to buy the organic tomatoes? If I consider the impact on the whole, on my own health and the health of the whole system, and if I have the dollar, then yes, I do. Do I spend the time and effort to grow tomatoes of my own? If I were to pay myself for the hours I spend gardening, account for all the money and effort and thought I expend, each tomato probably costs me thirty dollars (or more if I decide to raise my hourly rate)—terrible value for the money. But if I'm looking at more than the tomatoes, at the whole of what I need and value and take pleasure in—the value of the fertile, healthy soil I cultivate in order to grow tomatoes, the seven-year-old who lives in our house and likes to pick them off the vine (and the introduction it gives her to nature and the garden), the joy the bees take in the borage that grows with the tomatoes and the fruit the bees pollinate, the positive relationship with my friend Brook, who adores the green-tomato chutney I make, the uncountable value of eating something I have a real relationship with—then growing tomatoes is obviously of great benefit to the whole.

There are many small ways we can bring our daily lives into greater balance with our earth-centered values, from recycling our garbage to growing our own herbs for rituals. The hundreds of consumer choices we make are each an opportunity for affecting the greater balance of the whole. But there are two errors we can fall into around consumer choices.

The first error is becoming obsessive purists. In an imbalanced society, there is no way any one of us can become utterly pure. Today my computer is powered by the sun. I've traded in my pickup for an older, diesel model that can run on modified vegetable oil. I compost my garbage, grow worms, and horde every drop of water in the summertime. But I also fly on airplanes enough that I consume far more than an equitable share of the world's resources. I do that because there is no other practical way I can do the work I'm called to do, and I'm arrogant enough to think that the work is important, and justifies the fuel expended.

Not everyone has the extra dollar for the organic tomatoes, or the time or space to garden. Bringing our lives into alignment with the earth should not become a burdensome, guilt-filled project, where we are constantly in an unshriven state of eco-sin. Instead, we can think of it as a gradual, joyful process, where we look for the choices we can make that will enhance our lives. If I walk to a meeting instead of driving, I can enjoy the sights along the way and my own increased health from the exercise. If I'm too tired or rushed one day to walk, I won't flagellate myself for driving. If I can't afford to replace *all* my lightbulbs with compact fluorescents, I can replace one now, and one more each time I get a little extra money.

Making small choices that align with our values is important. It helps give us a sense of integrity, and it gradually transforms the whole of our lives to be in better balance.

The second error we can make around consumer choices is believing that those individual choices are enough to change the world. We live in a system that is currently so destructive, with so many large-scale destructive self-reinforcing cycles at play, that only *collective* action to change the larger system can hope to stay the damage and restore health.

Let's look at a larger question from the perspective of our definitions and understandings. Let's take the issue of agriculture and bioengineered foods, as posed in Chapter Two. If we were to look at the question of what benefits the whole, as defined above, Monsanto's Roundup Ready seeds would be seen as a horrifying travesty. Roundup Ready seeds are genetically engineered to withstand the herbicide glyphosate (trade name, Roundup), so that that herbicide can be applied wholesale to kill everything *else* that might compete with the crop. Although glyphosate is marketed as "safe," it has been shown to cause cancer,[2] and its use destroys the living community within the soil that creates a healthy environment for growth.

If we were truly interested in benefiting the whole, we'd boycott such products and instead look at ways to further organic agriculture and local food supplies, to support small farms, to make land available to more people, to bring food production as close as possible to where food is consumed. We'd understand that the vast majority of the billions who go hungry in this world are deprived not because there isn't enough food for them, but because they lack access to it or money to buy it. Before supporting policies that concentrate wealth in the hands of the few, we'd make sure that all people have what they need to thrive.

It's likely that policies like those outlined above would set off a new self-reinforcing cycle of benefits. Overall health would improve, from better-quality food and from a diminishment in pollutants and pesticides in the food and water supplies. Corporate profits would go down, but more real wealth and quality of life would be available to more people. City environments would improve, and small towns and rural areas would be revitalized.

We'll explore the issue of action toward the end of this book, after we have a firm grounding in the practices and insights of earth-based magic. But for now, let's look back at the question that opened this chapter, the question I asked the forest: How do systems change?

Systems change in response to forces that disturb their equilibrium. External forces, changes in conditions, new energies, and new challenges can shake up self-regulating cycles. So one way to change a system is to stir it up. That's the role of protest and direct action, and it's the reason why stronger forms of action are often necessary to bring change. Sweet reason, gentle persuasion, and dialogue that doesn't challenge the functioning of the system often end up becoming incorporated in the system's own efforts to maintain equilibrium.

Change in systems often comes from the edge. The edge, or ecotone—that place where one biological system meets another—is the most dynamic, most vulnerable, and often most diverse part of a system. The rocky shore where the ocean meets the land contains many more niches for life and diverse conditions for adaptation than either the sand dunes inland or the deep sea beyond.

So another way to change a system is to confront it with a different system. The existence of a feminist and earth-based spirituality movement offering rituals, teaching, and community completely outside the bounds of Christianity and Judaism has had profound effects on those religions over the past decades, offering support for reforms, challenges to established assumptions and practices, and creative ideas that have influenced change within the major denominations.

In spite of what the forest told me—that change has to come from outside because systems by their nature try to maintain themselves—I think systems *can* to some extent change from within. I'm not suggesting that every reader quit her job and go live in the woods. We are all part of the whole of the system, and to some extent that opens communication and makes it possible for us to influence it. To change a drum rhythm in a group of drummers, you first have to match it and join with it.

But when you are within a system, part of the whole, that system is also changing you. It is difficult to maintain your own rhythm and not simply become part of what you are trying to change.

Decades ago, feminist philosopher Mary Daly suggested that the place for feminists in the academy or other institutions was on the boundaries, neither completely within nor completely without.[3] Wherever we are, we can look for those fertile edges of systems, those places where unusual niches and dynamic forces can be found, and make change there.

Donella Meadows wrote a powerful essay many years ago entitled "Places to Intervene in a System," which detailed nine "leverage points" in increasing order of effectiveness.[4] The first two places to intervene, changing amounts and changing material stocks and flows, involve change in matter or form. If the school system is dreadful, pour more money into it or build new buildings. Sometimes those changes may be just what is needed, but they don't change the basic functioning of the system itself.

Next come changes in energy flow, looking at self-regulating and self-reinforcing systems and finding ways to intervene, either to disturb a nonfunctional equilibrium or to establish a new equilibrium to avert a crash.

Then come changes that begin to move into the realm of consciousness—changes in information flow, in rules, in self-organization, and in goals. Finally, the most overarching change comes from paradigm shift, a change in the basic premises that underlie the system.

We are faced today, in a world of global crisis, with the need for overarching change that can come only from a shift in paradigms, in our basic assumptions about the world. To change a paradigm, we must be able to express clearly what the new paradigm is.

That is the work of this book: to root us so firmly in earth that we can be walking emissaries of a new whole.

A SACRED INTENTION

Sit in a quiet spot and relax. You might want to meditate on the questions that follow, or journal about them.

Ask yourself, What is sacred to me? What do I care about so strongly that I can't bear to see it compromised or destroyed? What would I take a stand for? Risk myself for?

When you know the answer, consider for a moment what the world would be like if our social, political, and economic systems all cherished what is most sacred to you. In what ways do they already? In what ways would they need to change? What would change, in your daily life? In your community? In the world around you?

Can you describe that world in a few sentences or paragraphs?

Do you want to bring that world into being? Do you feel responsible toward it? If so, that is your sacred intention.

If not, what is your intention for your life? What are your goals?

Now consider how you spend your time and energies. Are your best energies directed toward bringing about your cherished vision of the world? Toward service of what is sacred to you?

If so, congratulations. Is there anything you need, support or opportunities or luck, to help you in that work? Who can you ask for support, in the human world? In the larger realms of the universe?

If not, what is blocking you? How would your life change if you were to put your best energies toward creating a world that cherishes what is sacred to you?

What do you need to make that change? Support? Opportunity? Courage? Luck? Who can you ask for support, in the human world? In the larger realms of the universe?

If you desire that change, affirm your sacred intention. Say, "It is my sacred intention to create a world that cherishes ________________."

Your sacred intention is the heart of your work with this book. You can revisit it and revise it, let it grow and develop, write about it in your journal, and test your daily decisions against it. As we go through this book, we will refer back to it again and again, as a touchstone for the transformations we undergo.

Rereading this exercise, I realize there's a deep assumption in it: that you, the reader, do feel a sense of responsibility for creating a world in alignment with your sacred values. Whatever you cherish is *part* of the earth, and loving it, wanting it to continue, to me logically implies a need to love and cherish the *whole*. I confess that for me that assumption is the heart of earth-based spirituality in this time when every life system on the planet is under assault. What is sacred to you may be different from what is sacred to me, and the ways you choose to serve it may not be mine, nor mine yours. But no one who loves the earth can evade responsibility right now for her well-being.

That responsibility may seem overwhelming. At times we fail to rise to it because we don't know what to do, or feel powerless or inadequate. I think it's no accident that *The Lord of the Rings* has become a highly popular series of movies at this moment in time. We are all Frodo, reluctantly carrying the burden of the ring, and not knowing clearly how we will get to our destination.

But as Galadriel, the wise elf, tells Frodo, "Even the smallest person can make a difference." Each time we act in service of our sacred intention, each time we align our energies and our actions with what we most truly love, we gain in personal power and ability. The path before us becomes clearer, and the help and allies we need come to us.

So, when you think about your intention, when you feel daunted or overwhelmed or afraid, just breathe deep and ask for help. Great powers and energies are all around us, but they cannot help us unless we ask. When we do ask, however, they are present and eager to help us serve intentions that benefit life. So you might say something like the following.

OPENING TO HELP

Great powers of creation and transformation in the universe, ancestors, allies, all beings who love the diverse and beautiful dance of life, I am open to your help and I reach out to you. I thank you for the gift of life, for the help and support I have already received, and for the great opportunity of being alive at this crucial moment. I need ____________________ to serve my sacred intention of creating a world that cherishes ____________________. I give you my gratitude for the help I know is already coming. Blessed be.

FOUR

Creation

What Every Pagan Should Know About Evolution

I managed to slide through something like nineteen years of formal education learning remarkably little science. In part, I was discouraged by a ninth-grade physics teacher whose experiments never worked. If she tried to demonstrate gravity, toy cars would refuse to roll down ramps and objects would float up. In later years, I majored in art, then film and psychology (which is science of a sort but didn't demand much grounding in biology or chemistry).

Now that I'm a Witch, I regret my ignorance and am taking steps to remedy it, mostly through reading and observation. The Goddess is embodied in the natural world, and science in its truest sense is about knowing nature. Thus our thealogy needs to be empirical as well as mystical.

Our understanding of our origins—cosmic and human—shapes our relationship to the world in subtle and profound ways. So hang on to your hats as we take a journey through the wonderful world of evolution, a topic that has always had profound religious and spiritual implications.

Most of us were raised on either the biblical creation myth or Darwin's theory—or perhaps both. From the Pagan perspective, neither of these stories is wholly satisfying or "true" (in the sense of best describing the reality around us).

The biblical creation story has a (presumed) male God making the world essentially by fiat, by word alone. The process is disembodied, entirely removed from the sweaty, bloody processes by which females create life. God's law is something imposed on nature, and God's rules are imposed on us to follow. Humans are made in God's image, and a great spiritual and existential gulf separates us from the animals. Plants, animals, and human beings were created in their finished and final forms, and have remained essentially unchanged since.

Of course, this view does not do justice to the breadth and diversity of Bible-based theology. There are strands within Christianity, Judaism, Islam, and all the major religions that celebrate and honor creation and preach a relational view of the world.[1] See, for example, the work of Matthew Fox in developing creation-centered spirituality, or Arthur Waskow's work on earth-centered Jewish ritual.[2]

Evolution, of course, was in Darwin's day a shattering and heretical challenge to the simplistic, literalistic biblical view. First, the theory of evolution holds that the world is much, much older than the Bible says. Second, humans, animals, plants, bacteria, and all other creatures are a single continuum of life. Humans are not something set apart. We are animals, and we emerged from the same natural processes by which other life-forms evolved.

From the perspective of earth-based spirituality, those insights were a vast improvement over literalistic interpretations of the Bible. Evolution restored dynamism to the universe, brought it alive as a growing, changing, interacting web of relationships.

Darwin himself was a great observer, embodying the permacultural principle of "thoughtful and protracted observation" more than a century before permaculture was formulated. He looked at the plants and animals and birds around him in the far-flung places of the world as he traveled, and he let himself ask, "I wonder . . . ": "I wonder how that tortoise got to be the way it is, how differences between those similar plants arose, what forces produced the beak on that bird." He theorized that environmental pressures and constraints select the individuals most fitted to a given environment from a range of genetic variations. Those individuals succeed best in the competition for food and scarce resources. They are also most likely to reproduce, and so they pass on their adaptations. His theory of evolution and natural selection was a brilliant example of relational thinking, focusing not just on individuals or species as separate, isolated elements, but on the whole pattern of interactions, exchanges, and effects of living communities as a whole.

But at the same time that Darwin was researching and writing, industrial capitalism was growing and consolidating its power, and looking for an ideology to justify ruthless exploitation of the poor by the rich. "Social Darwinism," a simplistic reformulation of Darwin's theory, turned natural selection into "survival of the fittest." The best win out and, by extension, the "winners" must be the best—and therefore deserving of their rewards. "Losers" are by definition inferior, maladapted, and deserving of their demise. To suggest that the winners owe anything to the losers is to interfere with nature and risk weakening the race.

This misinterpretation of Darwin's theory was a secular reformulation of earlier religious doctrines of the "elect." It was also a perfect rationale for cutthroat capitalism, in both the nineteenth century and the Reagan/Bush era. Competition is the driving force of progress in nature and, by extension, human society. The more worthy win out in time, and this, in the long run, is good for the species and for the whole. Success is its own justification, and what's good for big transnationals is good for the U.S.A.

There is a different view of evolution, one that better serves the worldview of earth-based spirituality. We might call it Gaian evolution, after the Gaia theory developed by James Lovelock and Lynn Margulis.[3] Gaian evolution is not so much a counter to Darwin as a shift in focus from the individual to the ecosystem, the whole. The earth functions like a living being, and the biosphere, the world community of life-forms, changes its environment as it is changed by it. The redwood tree does not evolve as a separate species; rather, the forest as a whole evolves, the interwoven lives of redwood and tanoak, huckleberry and salal, the mycorrhizal fungi in the soil below and the lichens in the canopy where the marbled murrelets nest. None of these creatures adapts alone, in isolation from each other—they coevolve as Forest-Being, in an interdependent dance that balances competition and cooperation. Individuals and species survive when their activities benefit the whole as well as the parts. Evolution becomes the story of how the planet herself comes alive.

Of course, scientists are very careful not to imply that this living planet has consciousness or self-awareness. Consciousness is not necessary to explain this process of life and evolution, and this becomes a messy and unnecessary part of the theory. With or without attributing consciousness or awareness to Gaia, we can still approach her life story with wonder and awe. For Witches, Pagans, and the like, however, having already removed ourselves from the realms of academic respectability, there are no reputations to protect, and thus we are free to experience Gaia as more than mechanistically alive—as a conscious being, a vast ocean of awareness in which we swim, always communicating, always present.

What follows is my synthesis of the story of Gaia coming alive, with thanks to James Lovelock and Lynn Margulis, and Elisabet Sahtouris's lovely book *EarthDance: Living Systems in Evolution.*[4]

Genesis

Before the Beginning. . .

In a swirling spiral of gas, heat, and light, a tiny grain of dust that was Gaia's seed danced and swirled. Throbbing and pulsing with an electric passion, she drew to her other grains, other seeds, until together they formed a ball, spinning and dancing in the lens of radiance that was to become the sun. The dancers flung out their arms, swirled their skirts, bumped up against each other, and fused. Growing larger and larger, spinning and dancing faster and faster, they were drawn toward each other by the passionate pull of gravity, at times colliding in a fiery death, at other times in a mating union, until at last the planets congealed into their orbits, circling a fiery sun.

Gaia was hot, her surface erupting in plumes and rivers of fire, her face bombarded by missiles of rock that left her pockmarked with craters and seeded with ice and the chemical prototypes of life. Slowly, slowly, she cooled down. On her surface, packets of energy frozen into form combined and recombined. Ice melted to primordial seas that washed a rocky shore. Lightning struck. Waves rolled to shore and retreated; the soup of energy was boiled and cooled, dried and immersed, again and again. Bubbles formed thin skins that enclosed crystalline strands of frozen energy, organized in a radically new way: a way that conveyed information, that communicated instructions for reproducing itself. The double helix of DNA was life's first great creative leap, the one that allowed all others to follow. Life was born.

The Gift of the Ancestors

Life on earth was still relatively new. At first, simple, one-celled beings filled the seas, living by changing the energy patterns around them, breaking down large molecules—complex clumps of dancing energy and form—into smaller clumps, using the energy released to move and dance. They filled the seas in promiscuous abundance, constantly exchanging bits of DNA, sidling up to one another and crooning the bacterial equivalent of "Hey, hey hey, baby . . . the thought of trading genes with you drives me c-crazy!" They formed one life-whole, one global gene pool, one planetary well of information and experimentation.

But after a time, life reached a crisis point. Life began to run out of food. There weren't enough of those complex molecules for all of life to continue, and life began to starve and die.

Yet life has always been inventive, creative. Those simple, one-celled beings were already experimenting with different forms. Some were long and skinny and wriggled and swam. Some were round and fat. Some adapted to hot and some to cold. And always they were trading genes, shifting forms, changing and transforming. At that time, there wasn't yet a brain on the planet, yet life came up with something so brilliant, so amazing, that it transformed the whole nature of existence and the atmosphere itself.

Life invented a mandala. A beautiful molecule, like a patterned flower, with a magic quality. For when a photon of sunlight struck the heart of this pattern, it began to vibrate and shiver and set off a chain of reactions that harvested the sun's energy to turn carbon dioxide and water into food. Chlorophyll and the process of photosynthesis were life's next great invention, and the green things, the sunlight-harvesters, were born.

Green things filled the seas and the crevices of the shores, flung a smear of filmy life over rock and sand. Life flourished as never before.

But there was one problem. The miraculous process that used sunlight to make food gave off a waste product, a toxic gas that burned and destroyed everything it touched. And as life grew, over hundreds of millions, a billion, then two billion years, the very air became polluted by this gas, so that life could no longer avoid its touch of death.

But life continued to experiment and invent. Some of those tiny creatures dug down in the mud to avoid the toxic gas. Some clumped together for protection.

And some discovered another miracle: that by reversing one of the moves in the dance of photosynthesis, a new process could be born—one that could take the toxic gas, which we call oxygen, and use it to burn food and make energy.

And so the breathers were born, those who dine on the sunlight-harvesters, burning their bodies as fuel for life. In burning food, the breathers give off carbon dioxide, which the green things (with the help of the sun) transform to food again. And the green things give off oxygen, which the breathers use in burning food. Gaia began to breathe, passing her breath back and forth from red to green, continuing to build up oxygen, to transform herself.

And after millions of years, the breathers took the mandala of chlorophyll, switched the atom at its heart to iron, and formed the hemoglobin that swims in the cells of our red blood.

And so the cycle is complete, and the earth breathes in and out, red to green to red.

And this air we breathe is a gift of the early ancestors. With each breath in, we take in the results of their great creativity. With each breath out, we give back.

And the balance is so perfectly kept that oxygen remains at just the right amount to sustain life. For if there were only a few more percentage points of oxygen in the air, any spark would light a fire that would ignite the whole atmosphere. And if there were only a few percentage points less, no fire would ever burn and we could not live.

Cooperation and Complexity

Breathers added something new to life's dance. The sunlight-harvesters floated in a womblike sea that contained the elements they needed to make food. The energy they needed showered down from the sun. Life was easy, and they could simply be and receive.

But breathers needed to find food, from the dead bodies of the sunlight-harvesters or from living ones. They had to be more mobile, more aggressive, pursuing and engulfing and penetrating before they could digest and dissolve their prey.

Every now and then, an aggressive breather penetrated a life-form that did not dissolve. Or took in someone it could not digest. And instead of eating each other, the life-forms coalesced and supported each other.

A breather might make food more efficiently for a scavenging bacterium. A sunlight-harvester and a breather might team up, to make best use of all possible sources of energy.

A long, skinny, wriggling creature might bury its head in this new, larger cell and provide mobility in exchange for food. A hundred, a thousand, tiny creatures might team up to become one larger being, pooling their crystalline DNA library of instructions into one central core.

And a new form of life was born, still single-celled but a thousand times larger than what had gone before, and far more complex. The *eukaryotes* were born, the cells with a nucleus that are the ancestors of all larger creatures.[5]

This new collective form opened up a wide realm of possibilities for life. For two billion years, simple bacteria had been the only model of life; now life began to experiment and change.

One of the first experiments was sex. Bacteria invented a simple form of sex, trading genes like bits of gossip throughout a worldwide pool. The variations created by this process allow them to change and evolve. When they reproduce, however, the process is still simple: each cell simply replicates itself and buds off an identical copy.

The eukaryotes each had a center, a nucleus that held a library of genetic information, arranged in paired chromosomes of DNA. Now they learned to divide those pairs, to split the deck before reshuffling. And each half-set of genes could combine with the half-set from a different individual. Sexual reproduction was born. And became very popular.

These new cells began to build on their cooperation, to form colonies. Some learned to take minerals from seawater and spin elaborate spheres of intricate, crystalline forms. Some pioneered the branching patterns of roots, the flat planes of leaves. Some built the first true bodies made of many cells, linked by the communication tubes of nerves.

And 580 million years ago, life exploded in variety. Life grew legs and began to walk on the seafloor and the shoreline. Life spun shells in elaborate, ornate forms, tried out tails, fins, flippers, segments, carapaces, and antennae. And this burgeoning life grew weirder, more delightful, more strange than all of succeeding life put together.

Until disaster hit. A meteor hit, or massive volcanoes belched smoke into the air, covering the earth with a blanket of cloud that first blocked the sun, then warmed the atmosphere. Something changed the earth's climate, and, in a great extinction, 90 percent of life on earth died.

What survived was not as diverse, not as inventive. But life tried out many variations on a few basic patterns and again began to grow and evolve. The seas were filled and the land was colonized. Dinosaurs roamed great forests of ferns, and winged lizards soared through the skies.

Then, some vast time later, another meteor crashed into the earth, leaving a great crater a hundred miles wide, and the dinosaurs died. Their small descendants, lizards and birds, still prowl and soar. A small, humble, shrewlike mammal that survived the cataclysm gave rise to mice and deer, tiger and mammoth, bison, horse, ape, and us.

And here we are, with our thumbs and our big brains, inventive, creative, aggressive, aware in some ways, oblivious in others, still struggling to learn the lessons our ancestors bequeathed us:

That everything changes. That everything is interdependent. That we survive by cooperating, sharing resources, pooling information. That change can come suddenly, cataclysmically, and when it does, the small are better fit to survive than the large. That when faced with great crisis, life is capable of great invention. That we are no less creative than the crystals that invented DNA, no less artists than the bacteria that shaped chlorophyll.

* * *

This is the story I like to remember when the world seems bleak and I wonder how we will ever survive. Or when I feel as if the Goddess and all powers of hope have deserted us. At those times I remember . . .

That life by its very nature is a great power of creativity and transformation, a power that will prevail.

And when I doubt, all I need to do is take a breath, in and out, and receive the gift of the ancestors.

Lessons of Evolution

What does this story teach us?

The first lesson is that the universe is amazingly creative, responsive, and ever-changing.

The second lesson is that survival goes not to the most ruthless or competitive, but to those who most effectively cooperate, communicate, and share resources. The cells of our body are collectives. The atmosphere we breathe is a collaborative endeavor. Adaptation is not about one species triumphing over another, but about a whole system coevolving.

Competition and predation are important aspects of this system, but they are not the only driving force. Cooperation and competition exist in a dance, a yin/yang harmonic balance of variation and culling, combining and editing.

The third lesson is the importance of variation, one of Darwin's insights. Life does not evolve one best type, but a multiplicity of fit types that retain a variety of forms and potentials. This variation, along with the random mutation of genes, allows for more diversity and more resilience in the face of environmental changes. A disease that fells one member of a species, for example, may not prove lethal to another. When a new potential food source appears, individuals who share a particular variation may be better suited than others to make use of it. Variation is every species' hidden treasure and insurance policy, and that is one reason why the loss of biodiversity and variation within species causes environmentalists such grave concern.

The fourth lesson is that change is not just a gradual process, but may also be a "punctuated equilibrium," involving periods of relative stasis punctuated by crises. The river of change generally flows gradually downhill along gentle slopes of slow, incremental change. But it may also plunge suddenly over a waterfall. Both forms of change occur in evolution and in life. When we look at potential environmental changes ahead, we do not know if we will see slow variations that our society and the biological community can adapt to, or sudden crashes down a rocky dropoff that will happen so abruptly that we cannot turn aside or cushion the fall.

If we were to extrapolate these lessons to human society, they might lead to a set of values we could call "social Gaianism" as opposed to social Darwinism. Social Gaianism would acknowledge individual needs and self-interest but see them best served in systems of cooperation and mutual aid. Our representatives in Congress might invoke the lessons of social Gaianism to bolster public support for social programs, childcare, education, health care, and other means of sharing resources and support. We would understand that the system is only as healthy as its weakest member. Libraries, those repositories of knowledge and inherited instructions, would be sacred. Knowing our interdependence with nature, and knowing also that a stressed system can suddenly crash, we would move swiftly and decisively to shift to renewable sources of energy, to phase out fossil fuels, and to keep chemicals out of our living and life-created atmosphere. When international conflicts arose, we would identify our security and survival with our ability to negotiate and cooperate, not conquer.

The Gaian story of evolution shifts our focus when we look at nature. We open our eyes and look beyond each individual tree to the pattern of the whole. We honor the unseen creatures below and above as well as what we can see. We begin to look for patterns and relationships, not just isolated individuals. We know that diverse, resilient, complex systems are most likely to survive. And only what's good for the biosphere is truly good for the U.S.A.[6]

This story also gives us hope. Enormous creativity is embedded in our very cells. Resilience is the nature of living beings. We, with our complex brains, have the inherent ability to evolve in ways that can nurture and sustain the life patterns that surround us.

And if we don't use that ability wisely, there are always those gene-swapping bacteria to sip the cocktails of our wastes and, in another billion years or so, come up again with something new.

FIVE

Observation

From my journal:

Sitting still in one place doesn't come naturally to me. But with a hurt leg, now seems the time to try it. I take my morning tea out to the deck, sit still, and practice owl eyes, the widening of perspective. I open my ears to a sphere of sound. The sun is just climbing over the hill; rays of light pierce through the forest. Tendrils of sensation up my back tell me I'm cold. I explore the sensation—could be fire, could be pain, could be just stimulation. I could sink into it, expand my comfort zone—but then I'm a middle-aged woman susceptible to colds, so I go inside and get a poncho.

A loud voice—the Eight-Note Bird, eight sharp, slightly tart notes in a string. I know he's close by but I can't see him among the branches. Nearby is also a small-voiced bird that sings simply "Wheet, wheet" at intervals. Beyond I hear the wind in the branches and the trickle of water in the lower stream. Down here in the canyon it's still and windless, but I can look up and see the tops of the redwoods swaying, and I hear an occasional sharp, ominous creaking.

The light moves, the rays change their angles, and suddenly a spiderweb gleams iridescent before me, hanging in the branches of the trees, a perfect wheel. I look and see another, and another—invisible until the light strikes them at just the right angle. Then, for a few precious minutes, they glow in rainbow colors, gold within, an iridescent purple in the outer rim. The air is full of circling bugs who are also illuminated, spiraling upward toward the sun. Threads of light gleam where spiders have spun bridges between trees.

A phone call summons me indoors, and when I return everything has changed. The sun has cleared the hill and a chorus of birds begins to sing. While Wheet and Eight-Note sing the same words over and over, now I hear a bird with a song as complex as a paragraph, an essay of rhythm and melody that plays once, and then no more. Other birdsongs, more liquid and sweet, sound in the background. In the deep forest, I hear a dove call. The ravens announce their presence with a caw that is almost a honk, then cluck a few times before they fly by. A jay squawks; I hear a woodpecker shriek, shriek, shriek as it flies.

A scratching and rustling overhead—the branches of the redwoods spring back and I look up to see two gray squirrels chasing each other through the treetops. I follow their mad dash with my eyes: from directly overhead they run out along the redwood branches, leap to the tanoak, travel up and down the highway of treetops. Another phone call. This time when I'm inside I grab my binoculars, and sure enough, when I return the squirrels are back, graceful and daring as any high-wire artists. Through the binoculars, I can see one up close—his neat white belly and fluffy gray tail. He looks at me—I've noticed that when I use binoculars, animals and birds often seem to sense that they're being watched, that some boundary of distance has been crossed. He stands still on a stump of a redwood, staring at me in a posture of tension for a long time, then hears something behind him and turns.

The other night, after a forester we were consulting informed us that squirrels had caused the tops of a couple of our redwoods to die, I dreamed I was making love to a squirrel. In the dream, he looked like a man, but when I felt his face, it felt like a muzzle. "I could tell you have serious hands," he said. "We don't want that, now, do we?" Just another guy who can't make a commitment, *I thought, and then,* But what do I expect? He's a squirrel! *Now, watching the lithe bodies leap and bounce on the high branches, spying on the sleek, gray fellow, I think a squirrel lover wouldn't be so bad. Even nicer would be to run and leap among the high branches with that freedom. Rima the Bird Girl, Julia Butterfly, the Earth First! tree-sitters who play tag eighty feet above the ground—maybe they know that delight. Me, I'm earthbound, but happy watching the squirrels perform their high-wire acts for my amusement.*

The first skill that can lead us to deepened connection with the natural world is observation. Observation seems like the simplest, most natural thing in the world. We're all born observers. From the earliest moments of life, babies stare at, listen to, and taste the world. Parents of two-year-olds soon learn to watch their language, for their observant progeny will reliably mimic any favorite swearwords or verbal quirks. By observing others, we learn how to speak and what to say, how to behave and misbehave, how to experience and navigate the world we're born into.

But between infancy and adulthood, something gets in the way of our childlike, unclouded vision. Few of us can walk into a forest and simply *be* in the forest. Instead, by adulthood we are inside a story we're telling ourselves, partly about the forest, but mostly about ourselves. Sometimes it's a story about our own weakness and inadequacy: "I'm so tired. I don't feel well. I can't keep up; everyone else is in better shape than I am." Sometimes it's a story about how wonderful we are: "I'm so spiritual. I'm attuned to the trees so much more sensitively than anyone else. I'm talking to the faeries. Don't I look good here in the woods?" Sometimes it's a story about how much we have to do: "I have that phone call I forgot to return and I haven't checked email in two days and then I have to get the car fixed and the insurance premium is overdue." Sometimes it's a story about someone else: "I'm so angry! Why did she say that? I can't believe her—why did she lie about it? I'm going to tell her to her face just what I think of her. Better yet, I'm going to send an email out to the whole listserve for the group." Sometimes it's a story about fear: "What was that? I think we're lost. Was that a cougar? I've heard cougar attacks are increasing. What if we can't find our way back? What if it starts to rain? I've heard you can die of hypothermia even on a warm day. My heart is pounding. What if I have a heart attack? What if I slip and break an ankle here? What if there's a rapist lurking in these woods?"

Whatever inner dialogue we're running, it's interference. We end up walking around inside our own heads, not in the woods.

So to truly observe, we must be able to step outside our heads and walk out into the woods. We must be able to close the book on the story, turn off the dialogue (or at least turn it down), and hear what's around us.

Of course, that's easier said than done. Buddhists spend years in practice in order to achieve a quiet mind. Psychotherapists work with their patients for session after session in order to change their internal stories.

Paradoxically, observing the outer world around us requires a great deal of inner work and discipline. It becomes a deep spiritual practice that incorporates some of the aspects of Buddhist detachment and may lead us on a journey of personal healing.

Below are some of the exercises and practices I use.

OPEN-EYED GROUNDING

Grounding is the basic practice that begins every ritual, and the tool that helps us stay calm and present in any tense situation. Grounding means being relaxed but alert, energetically connected to the earth but able to move, present and aware, in a state in which we can take in information and make conscious choices about what to do.

In a ritual or meditation practice, we often close our eyes in order to ground. Closing our eyes can be a helpful way of focusing on our inner vision and shutting out distractions. But in the woods, or in truly dangerous situations, we may also want to be able to ground ourselves with our eyes open. It helps to practice.

You can do this exercise anywhere. Ideally, you should be outside, standing on the earth, in some natural place. But you can also do it indoors or in the midst of a city.

Stand in a comfortable position, with your feet about shoulder-width apart, your knees lightly bent. Stretch and release any tension. Take some long, deep breaths into your belly. Feel your feet on the ground.

Tell yourself that just by breathing and feeling your feet on the ground, you can bring yourself into a calm and grounded state.

Imagine that, like a tree, you can extend roots into the earth, from your feet and the base of your spine. With your eyes open, notice if what you see changes as you extend your roots down.

If there is anything clouding your awareness or interfering with your ability to be present, take a deep breath and imagine letting it go down through your roots into the earth, to become compost.

Feel the living fire deep in the heart of the earth. Breathe some of that energy up through your roots, into the base of your spine and your belly, up your spine as if your spine were the flexible trunk of a tree. Feel it warm your heart and throat, and reach out through your arms and hands. Let it move up through the top of your head and out like the branches and leaves of a tree. Let those branches come all the way down to touch the earth, surrounding and protecting you. Again, notice what changes.

Feel the sunlight (or moonlight or starlight) on your leaves and branches, and breathe it in. Feed yourself on the energy, just as a tree feeds on sun. Draw it down through your head, heart, hands, and belly, down through your feet into the earth.

Look around you and notice what you see, hear, smell, and feel in your grounded state.

EARTH-WALKING: MOVING WHILE GROUNDED

Now you are grounded: calm, present, and aware, with an energetic connection to the earth established. But, although we've used the image of a tree to help us ground, we don't want to be stuck to the ground or immobile. We need to be able to move while staying rooted.

So imagine that the roots in your feet are stretchable, that when you pick each foot up the connection remains. Or you might use the image of the Ents, the tree people from Tolkien's *Lord of the Rings,* who stalk the forests on their treelike feet, with their toes spreading and gripping the ground with each step. Looking up and around you, not down at your feet, begin to move, remembering to breathe.

Imagine that your feet have sensors. Pick each foot up and set it gently and slowly down, letting it tell you about the terrain beneath. Keep your ankles loose and your knees springy. Move slowly around the space, feeling how you can keep your energetic connection with each step.

Notice how quietly you can walk. Suddenly the squeak of your shoes or the swish of your clothing becomes loud. Practice moving in this quiet, grounded state.

WIDE AWARENESS

Now stop for a moment. Looking straight ahead, bring your arms out to your sides and wiggle your fingers. Slowly bring them in until you can just see the motion of your fingers with your peripheral vision. Notice how wide your field of vision can be, how far it can extend up and down as well.

Letting your arms drop down, keep your vision extended and again move through your space, earth-walking in a calm and grounded state. Notice how quietly you can move, how much you can see and be aware of, and how you feel, earth-walking in wide awareness.

The three linked exercises above—open-eyed grounding, earth-walking, and wide awareness—are the core of my own daily spiritual practice. The key is, indeed, to *practice* them—practice them daily, moment by moment, until they become second nature, in the woods or in any situation in your life. In tense or dangerous situations, you will be safest if you don't panic and instead stay grounded, calm, able to assess what's happening and make conscious choices about what to do. In fact, when I prepare activists for political actions, I teach a version of these same exercises.

For many of us, breathing and grounding requires some deep repatterning of our basic way of being in the world. Many of us have learned to breathe high in our chest and shallowly, which keeps us in a state of heightened anxiety. Our entire culture is one of distraction, keeping us focused on other people's stories—stories we see on TV or read about in the newspapers. If we have experienced trauma or violence in our lives, we may have learned to close down and draw a shell around us, to avoid seeing or feeling what is too painful to face.

Repatterning takes time. Be patient but consistent. Learning to ground does not require hours of strenuous practice every day; it can be done in minutes, in odd moments of time—while waiting for the bus or walking the dog. The best practice is to remember to do it in moments of stress, but remembering to ground in the midst of stress takes a lot of consistent work. But when you do, when you begin to automatically breathe and ground in response to tension, you will find that you can handle situations of high intensity much more effectively and easily than before. And you will be better able to be present in the woods, nurtured by their beauty.

In magical practice, we learn to awaken subtle senses that can let us perceive energies and spirits that go beyond the physical world. But too often, we focus on the nonphysical without first being fully rooted in an awakened, sensual experience of the physical world. Below is a simple exercise to help us awaken our senses and be fully present and aware in our bodies.

COMING INTO OUR SENSES

After grounding and practicing wide awareness, find yourself a safe and interesting place in the natural world—which could be a pristine spot in the wilderness, but could also be your backyard, a quiet corner of a city park, a vacant lot. Now close your eyes, just to shift your focus away from what for most of us is our dominant sense.

Sniff the air. Take some long, deep breaths through your nose, followed by some short sniffs. Become aware of what you smell. The air is full of information. Imagine for a moment that you have the nose of a dog or a wolf. What would the breeze be telling you? Can you smell the trees? The moisture on the wind? The chemical tang of polluted air?

Taste the air in the back of your throat. What are the tastes still lingering from breakfast? Does your saliva have a taste? Roll it over your tongue as if you were tasting a fine wine. What parts of your tongue come alive? What information does it give you about your state of being?

Feel the air on your skin. Become aware of the touch of the breeze, the temperature. Are you standing in the sun or the shade? Are there patches of both, and how do they feel different? Become aware of your weight, your stance, the pull of gravity on your body, your sense of balance or energy or fatigue.

Open your ears. Imagine that you have the ears of a deer, that you can shift and point in any direction. What sounds do you hear? Do you hear birds? Traffic? Voices? Insects? What do your ears tell you?

Now open your eyes. Add sight to the information you are receiving from all of your other senses. What do you see when you focus on a point or an object? What do you see when you extend your vision into wide awareness?

Earth-walk in your space, in wide awareness with your senses open. What do you smell and taste and feel and hear and see?

Being Rooted in a Place

As the storybook prince mentioned in Chapter One discovered on the Isle of Birds, learning to hear something takes time. To truly observe the patterns of the natural world, we also need to be rooted in a place. And our observations will be most clear and focused if that place is very specific, a special spot (or "home base") that we return to over and over again, ideally on a daily basis.

Some of my great teachers about observation and the natural world have been the folks at the Wilderness Awareness School, who teach tracking and nature awareness from a spiritual perspective. The core of their teaching is the importance of a "secret spot," a place that you return to regularly to look and listen, that you ultimately know intimately.

Of course it's wonderful if that spot can be out in the pristine wilderness, but because the home base needs to be somewhere that we can return to regularly, a wilderness spot isn't practical for most people. City dwellers do better to make it our backyard or the park across the street—a place we can get to easily in the course of our normal day's activities. Very few of us have hours every day to spend observing the natural world. Given the busy, stressed lives most of us lead, we're lucky if we can carve out some minutes here and there.

Cities can be fertile places for observing the natural world. I've seen a pileated woodpecker and a great snowy owl in the heart of inner-city San Francisco. Peregrine falcons nest among skyscrapers, and raccoons plunder the garbage cans in many neighborhoods. But besides the "wildlife" that exists in an urban setting, human beings are also animals and follow our own natural patterns. Economist Jane Jacobs is one of the great observers of the ecological patterns of city life. The following passage is a beautiful example of urban observation from her classic work *The Death and Life of Great American Cities:*

> The stretch of Hudson Street where I live is each day the scene of an intricate sidewalk ballet. I make my own first entrance into it a little after eight, when I put out the garbage can, surely a prosaic occupation but I enjoy my part, my

little clang, as droves of junior high school students walk by the center of the stage dropping candy wrappers. . . .

While I sweep up the wrappers I watch the other rituals of the morning: Mr. Halpert unlocking the laundry's handcart from its mooring to a cellar door, Joe Cornacchia's son-in-law stacking out the empty crates from the delicatessen, the barber bringing out his sidewalk folding chair, Mr. Goldstein arranging the coils of wire which proclaim the hardware store is open, the three-year-old with a toy mandolin on the stoop, the vantage point from which he is learning the English his mother cannot speak. Now the primary children, heading for St. Luke's, dribble through to the south; the children for St. Veronica's cross, heading to the west, and the children for P.S. 41, heading toward the east. Two new entrances are being made from the wings: well-dressed and even elegant women and men with briefcases emerge from doorways and side streets. Most of these are heading for the bus and subways, but some hover on the curbs, stopping taxis which have miraculously appeared at the right moment, for the taxis are part of a wider morning ritual: having dropped passengers from midtown in the downtown financial district, they are now bringing downtowners up to midtown. Simultaneously, numbers of women in housedresses have emerged and as they crisscross with one another they pause for quick conversations that sound with either laughter or joint indignation, never, it seems, anything in between. It is time for me to head to work, too, and I exchange my ritual farewell with Mr. Lofaro, the short, thick-bodied, white-aproned fruit man who stands outside his doorway a little up the street, his arms folded, his feet planted, looking solid as earth itself. We nod; we each glance quickly up and down the street, then look back to each other and smile. We have done this many a morning for more than ten years, and we both know what it means: All is well.[1]

A HOME BASE

Find a spot that will be your home base, where you can practice the magical disciplines of observation. Be modest and realistic in your expectations of what you can do. Is there a park or a community garden or a vacant lot full of weeds near your workplace where you can eat your lunch regularly? Is there an untended hillside you walk by when taking your child to the playground or your dog for her daily walk? Do you have a deck where you can drink your morning coffee and look out at the garden?

Once you have your spot, spend time there. Do the awareness exercises offered above, or just sit and listen. Notice how what you observe changes over time. How are the birdsongs different through the seasons? What animals or human patterns do you observe, and are they different at different times of day? In different weather? On workdays or holidays?

And how does your own mental and emotional state affect your ability to observe?

Letting the Stories Go

I am not an obvious candidate for the role of wilderness instructor. Throughout my life, I've consistently been the last person in any group going uphill, the slowest person on every hike. I was not one of those bold, physical children who are constantly testing their abilities and challenging themselves to run, jump, climb, and do daring things. I liked to read books. Left to myself, I would have spent my entire childhood curled up in my bedroom with a good fantasy story. My mother had to insist that I go out and walk at least six blocks a day in the summertime.

As a result, it's really easy for me to slip into a story in the woods about how weak and out of shape I am, especially when I'm in a group of faster, stronger people. Instead of being in the woods, I can easily be in some internal proving ground where my sense of self-worth is on the line—and not being confirmed.

I've had to learn to let that story go. It's a constant practice, and I'm never totally finished with it. In one of my novels, *Walking to Mercury*, the character Maya Greenwood learns to let her story go while hiking up the Himalayas with a bad cough. Fiction is different from autobiography, but this incident was based on a real moment for me.

> She sighed. The path continued up and up, with no relief in sight, and she really couldn't sit down on the trail and cry. She had to go on. There was no choice. . . .
>
> All right. Here she was in Buddha land. Why not try the Buddha path . . . and let go. Let her body be, with its limitations and imperfections, its wheezing lungs and its extra pounds of flesh. Stop trying to carve it and chisel it and make it into something it isn't, and love it like a pine, like a boulder. Stop blaming herself for being sick and slow and heavy. Let go of the words *sick* and *slow* and the weight of all they conveyed, and just feel the workings of her muscles against the rock. Stop wishing the uphill were downhill; stop telling herself a story about what she should feel, and let be.
>
> She stopped. Just for a moment she let herself breathe in the clean, thin air. She was surrounded by incredible beauty; white forms on the blue crowns of mountains so high they appeared just where you'd expect only birds or stars

> to be. For that one moment, she was present in the beauty, no longer wishing she were somewhere else. . . . The path was still steep, her breath still rasping, her body still slow and terribly tired, she was still worried . . . but somehow none of that was between her and the mountains any longer. She was released from a glass cage full of chattering, clamoring noise, into a world where she could feel the air on her skin and hear the bell-like tones of silence. Light danced off the glaciers to caress her eyes, and she opened, letting herself be emptied, disemboweled, a conduit for wave after wave of love deep enough to match the mountains.[2]

Three years ago I broke my ankle on a hike through the woods that was part of a goodbye party I was throwing myself before going off to Europe for three months of teaching and lectures. I slipped on the gravel of a dirt road and landed wrong on it. Instead of enjoying a barbecue with all of my friends, I ended up in the hospital having surgery. I spent the next week at home, recovering, waiting for the ankle to be ready for a cast, grumbling at my partner and housemates and occasionally tossing the odd piece of furniture around the house, while snapping viciously at the parade of well-meaning friends trying to assure me that this was the Goddess's way of telling me to slow down. I pointed out to them that if I had a boyfriend who thought I should slow down, and communicated that thought by breaking my ankle, they'd be the first ones urging me to get out of that abusive relationship and report him to the relevant authorities. I give the Goddess every opportunity to communicate with me on a daily basis, I assured them, and if she wanted me to slow down, all she had to do was say so and reduce the workload a bit, not break my ankle, which in any case didn't slow me down much. I went off to Europe with crutches and a wheelchair a week later.

Throughout the trip, I could feel stories hovering. Being in a wheelchair and dependent on others for many things I could normally do myself wasn't easy. I could hear the whispers of the stories that shaped my childhood—stories my mother told herself about being a victim, being abandoned, never being taken care of, always caring for others, and never getting enough herself. Stories I had internalized, as well.

But I also became aware of something else. When I got caught in the stories, I felt abused and misused and unhappy. When I could stay out of those stories, being in a wheelchair with a broken ankle was just something new I was dealing with. It was interesting, actually—a different sort of a trip than I had planned on or hoped for, but a trip that showed me a whole new perspective on the world that I wouldn't otherwise have seen. I had a whole new relationship to sidewalks and curb cuts and bathrooms. I understood the need for laws protecting access for the disabled in a whole new, visceral way. I look back on it now as an extremely valuable experience.

How do we let the stories go? Some of us need deep emotional support to do so, in some long-term form of counseling or psychotherapy. Twelve-step programs to help people recover from addictions are another healing discipline that can help us reshape our lives as well as our ability to be present in the woods. And a supportive ritual circle or coven can also help in our personal transformation.

The exercises above and below can help sharpen our inner observation and ability to make choices, and can work independently or in conjunction with counseling or a support group. Ultimately, magical discipline is about learning to hear, understand, and "speak" to our own inner states of consciousness as well as the outer world: the goal is to become aware of our current state of awareness and to make conscious choices about what state we want to be in. And every form of personal change requires our active will and participation.

SELF-OBSERVATION

Sit in your home base or some other safe and quiet space. Begin to practice the awareness exercises. Now notice what gets in the way. Ask yourself,

What internal dialogue do I have going on in my head? What story does it represent?
What energy does it bring with it? Does it feed my energy or drain it?
What emotions am I feeling? What is my physical body telling me? What muscles are tight? How am I breathing?

WHAT CHARACTER ARE YOU?

Once you have a sense of your story, or of the emotions and energies constricting you, ask yourself, "If I could name the character I play in my own story, what would she/he be called?" Does your character come with a favorite phrase or bit of characteristic dialogue? Sometimes I'm Ilse the Fascist Healer, who says, "You *will* get better . . . or else!" One of my students came up with Bitter Betty, who says, "I take care of everyone else, but no one takes care of me!"—the exact words I heard from my own mother, over and over again. Or I might be Frantic Frannie, who says, "I have more work to do than I can possibly do in the time I have, so let me do five things at once!"

Now consider *your* character. You might want to do some writing in a journal about her/him. If you're working in a circle or support group, your friends might also help you consider some of the following questions:

In what ways is this character's experience of the world narrower or more constricted than it might be?
Are there ways in which this character expands my experience of the world, or serves me?
When I observe through this character's eyes, what do I not see or do?
How does this character influence the choices and decisions I make?
How would my experience change if I were a different character?
What do I want to do with this character? Tell it to go away? Kill it? Love it? Absorb and integrate it? Recognize and laugh at it?

Some of the worst mistakes in judgment I've ever made have been under Ilse's influence. Now, whenever I'm engaged in healing work, I take a moment to consciously acknowledge her and send her away, to make sure that my choices are coming from some other inner state. I've learned that I can do an enormous amount of work as long as I'm *not* being Frantic Frannie; but as soon as she starts in I get frazzled, fried, and exhausted, and in fact become far less effective in what I do. I miss the freeway offramp because I'm trying to talk on the cell phone while driving to save a few moments and end up going miles out of my way, getting caught in rush-hour traffic, becoming mad and frustrated, and wasting time. When I hear Bitter Betty's voice, I know that I have to stop whatever I'm doing and rest or do some self-care, or else I will get sick.

As Ilse, my ability to observe what's going on with someone who needs healing is colored by my own need to be a savior. I cannot clearly take in information. As Frantic Frannie, I can miss even clear information that's right in front of my eyes: a huge sign saying, "San Rafael Bay Bridge, Next Exit." As Bitter Betty, I can't see or take in the many ways my friends and partner are nurturing and taking care of me. Still less can I take in the healing and nurturing of nature.

A group or circle can help us to identify these characters, and this work can also be powerful and healing for the group. When we recognize our own constricting characters and introduce them to each other, we can all laugh at them together. Knowing that my circlemate sees and acknowledges her own irritating side makes me feel more forgiving and tolerant. Knowing that she sees and accepts my own annoying personality traits makes me feel freer to be who I am, and more fully accepted and deeply loved. When I feel loved, my own love for the earth can flow freely.

Anchoring

To release those constricting characters, it's vital that we know what it feels like to be without them, to be as unimpeded as an animal on its home turf, loping along a familiar trail in its accustomed gait. We want to be in a state where we are relaxed, somewhat neutral—not agitated or putting out energy, but with energy and power available to us. This state, which we call the "core self" or "baseline state," is so valuable that I teach a magical tool called "anchoring" to help us identify it and return to it quickly.[3]

Achieving the baseline state is more than just being grounded. "Grounding" means having an energetic connection to the earth, being present, aware, and in your body. Being at baseline is all of that, but it also implies being emotionally at neutral, not inflated or pumped up, not depressed, not telling yourself a story about yourself.

Neutral is not numb. Rather, it's the state of being maximally open to the information and communications coming to you. It's the place from which you can quickly shift into a different gear, if a need arises for action or speed.

"Anchoring" is a tool to help us reach a particular state of consciousness—*any* state, including being at our baseline. It means associating a particular state of consciousness with a physical touch or posture, a visual image, and a word or phrase. With practice, we can use that anchor—that combination of touch, word, and image—to bring us quickly into the desired state of awareness. With further practice, any one of the three may be enough.

We could create an anchor, for example, to help us reach that state we call being grounded—relaxed, aware, and energetically connected to the earth. We could create a different anchor to help us move quickly into a deep trance state in which our awareness is focused internally. In the exercise below, we create an anchor to help us move into our baseline, neutral state.

ANCHORING TO CORE SELF/BASELINE

Begin in your home base or another safe space. Ground and center yourself. This time close your eyes. Think of a place or time or situation in your life where you feel relaxed and at ease, where you can just be yourself. Where you don't have to impress anyone or achieve anything or exert your power, but where you have power and energy available to you if you need them. Where your focus can be on the world around you, not on yourself.

Say your name to yourself, the name you most identify with. Notice where it resonates in your body, and touch that place, or find a posture or gesture you can make that feels connected to this baseline state.

What image or symbol or picture comes to mind that can embody this state? Hold it in your mind as you also touch that place in your body or make your gesture.

Is there a word or phrase that comes to mind, a magic word you can use that calls you into this state, a phrase that can counter the phrases of your constricting character? Say it to yourself as you visualize your image and make your gesture.

When you use these three things together—the physical gesture/touch, the word or phrase, and the image—you can bring yourself instantly into this grounded, neutral, core state.

Now go through the exercise of coming into your senses, noticing what you can smell, taste, feel, and hear. Open your eyes and notice what you see.

How does your ability to observe change in this state?

Use your anchor and practice coming into this state regularly. Add it to your daily practice of grounding, and use it whenever you go to your home base to observe.

And practice using it when you get caught up in the dialogue or vision of one of your constricting characters. Notice what changes as you shift away from a constricted state into an open, grounded, neutral place.

As with grounding, the more you practice with your anchor, especially in moments of tension and stress, the more automatic it will become, until eventually for you it becomes normal in moments of stress to ground and go to your core self. From that inner place, you will be better able not just to take in information but to make conscious choices, to act instead of react.

I have several personal anchors that I use to move in and out of states of consciousness quickly. For example, one summer when I was traveling alone in southern France, visiting the ancient caves filled with art of the old Stone Age, I very much wanted to be able to meditate in those sacred places. But the only way to get in was to go on a guided tour, which did not allow for long, silent trances in the dark. I used a simple gesture so that I could drop into a state of deep meditation while walking, and could listen for the voices of the ancestors. But be careful: when you are anchored to a state of deep meditation, you are more open and vulnerable than usual. One day the ticket-taker on the tour accused me falsely of not paying for my ticket, and I was surprised at how shaken up I felt, and how much difficulty I had explaining myself and remembering my French vocabulary—until I remembered that I was in trance.

The neutral or baseline state, however, is probably the most useful state of consciousness for me. I anchor to that state when I'm facing a difficult or dangerous situation—because when I am neutral I am most capable of taking in information. I walk in the woods in my baseline state, so that less of me and my story gets in the way of what the forest is showing and telling me. I anchor to neutral when people are praising me effusively, or damning me thoroughly, so that I don't get my own core worth confused with praise or blame.

I Have To / I Choose To

The stories and characters that constrict us limit our choices. On a personal level, they keep us acting in set patterns instead of freely choosing what we want to do. We repeat the same negative patterns in relationships or work, stay stuck in bad situations, make the same mistakes over and over. On a collective level, we often rationalize and justify our perpetuation of negative patterns in the same way: we justify abuse and violence by saying, "We have no choice."

"What can we do—we have to do *something*," my neighbor says when I ask him whether he thinks bombing Afghanistan is the proper response to 9/11. "I don't want to be doing this," a soldier who is detaining civilians tells me at a checkpoint in the West Bank of occupied Palestine. "But what can we do—we have no choice." "I don't want to step on you with my horse," a policeman said to the young woman next to me when we were sitting in front of a line of mounted police in a demonstration. "But if my captain orders me to, I won't have a choice."

The essence of nonviolence is choice: acknowledging that we always do have a choice about whether or not to use violence, and posing an expanded range of choices to those in power. In fact, we had a broad range of choices we as a society could have made in response to 9/11. The soldier at the checkpoint could have chosen—and eventually did choose—to let the group of men he was detaining go home to their village. No court in the world would have convicted the police officer on the horse for refusing an order to trample a young woman—and, in fact, the order never came and the police instead withdrew.

It was clear to me, in each case above, that the authorities were saying "I have to" in order to absolve themselves of responsibility, to avoid choice. But as I thought about those incidents, I began to become uncomfortably aware of how often I said "I have to" to myself. I had to be sitting in front of those horses, because—well, because I just felt I *had* to be. If I were to be trampled or hurt—well, I had no choice. I *had* to be there. It would not be my responsibility.

But what would change, I began to wonder, if instead of "I have to be here," I told myself, "I *choose* to be here."

Saying "I have to" disclaims both responsibility and credit. It opens the door to Bitter Betty and her self-pitying, victimized sisters.

In saying "I choose to," I claim my power. Why would I choose to sit down in front of a line of horses? Because for twenty years I'd been telling people in nonviolence trainings that that's how you stop horses, and I wanted to stop that line from advancing on the crowd. Intuitively I felt that if the line kept advancing, someone would get seriously hurt, and I was willing to take a personal risk of getting hurt in order to prevent that. If I did get hurt (and fortunately I didn't), I would not be a passive victim, but a person who had made a choice and had accepted the risks that went with it.

"I choose to" also opens the door to a wider set of questions. Before I sit down, I might ask, "Am I willing to take this risk? Is this the moment to make a stand? Is it worth it? Can the horses and their riders see us? Are there constraints on the violence they might inflict upon us? Am I choosing to do this from a grounded, anchored place? Is this truly my task?"

When we take responsibility for healing the earth and restoring the balance, we are faced with many choices. Some may involve what seems like sacrifice. No, I won't buy that new SUV. I'll walk to class instead of driving. I'll take the trouble to recycle that can. Some choices may involve situations of discomfort or danger: I'll sit in front of those horses because I'm protesting an institution that sets policies that destroy the earth and human lives.

Whatever choices we face, it's important that we stay in our own power as choosers, that we don't see ourselves as martyrs or victims, but as active agents exercising our freedom and our will. Then we remain empowered, whatever consequences we face, and are able to act joyfully, courageously, and creatively.

So observe your own inner dialogue. Whenever you find yourself saying "I have to," stop, ground, and use your anchor to your core self.

Then ask, "What would change if I said 'I *choose* to'?" What new questions would arise? What would be the basis of your choice? What responsibility would you acknowledge? What greater credit might you claim?

Frantic Frannie has to go to a meeting instead of the movie she's been wanting to see, and she sits there resentful and sighing, running out at crucial moments to make a call on a cell phone, interrupting discussion to ask when the meeting will be over. She doesn't actually contribute much to or get much out of the meeting, and the tension she generates makes the meeting less productive for everyone. Rather than letting Frannie dictate my actions, I might instead choose, from my core, grounded self, to go to the meeting instead of the movie, because it's a key part of something I care deeply about. Having chosen to be there, I'll want it to be as productive and fun as possible. I might bake an apple pie to bring, or offer to facilitate. I'll be glad to see the people I know and enjoy working with, and my gladness will generate a positive, good feeling in the room. I'll listen attentively and maybe have helpful comments to make or inspiration to share.

I don't believe that *everything* in our lives is a matter of choice. In New Age circles, I often hear people say, "We create our own reality." That's a shortsighted and simplistic misunderstanding of how reality works. We don't choose all of our circumstances, or our range of choices. The poor don't generally choose to starve, nor do the oppressed choose their oppression, and the casualties of war don't choose to die.

But we can choose how we respond to the circumstances we're presented with. I didn't choose to break my ankle, but I did choose to go on with my trip anyway, to enjoy it as much as I could, to accept that I'd have moments of anger and frustration while striving to stay open to a new mode of experiencing the world.

All this may seem to have strayed far beyond the woods and the practice of observation. But it is one of the wonderful paradoxes of magic that everything works in circles. Outer work leads us around to inner work, and inner work allows us to do the outer work.

Bitter Betty can't enjoy the woods because she is never truly in them; she's walking around inside her own replay of that nasty remark that Sullen Susan made about her that she simply *has* to respond to. Frantic Frannie can't enjoy the woods because she *has* to write this article and make that phone call and answer that email, and she doesn't have time.

But I can enjoy the woods, from my grounded, core self, because I choose to make time to do so, and to be fully present and aware when I'm there.

KEEPING A JOURNAL

One of the myths of Witchcraft is that every Witch of ancient times kept a "Book of Shadows," a magical journal that recorded her spells and charms and herbal recipes. In reality, most Witches of that day were probably illiterate, but a Book of Shadows is still a good idea, especially when we begin observing and learning from nature. Taking time to record your observations will help them become more clear to you, and over the years those observations can prove an invaluable record. What did the birds sound like at this time last year? Are there really fewer of them, or does it just seem so? What were the mushrooms I found on my walk last year? Is the weather really different? Keeping notes in your journal will help you answer such questions, and provide a record of how your own abilities to hear and understand grow and develop.

NAMING AND IDENTIFYING

Learning the names of trees and birds and mushrooms may seem like a stuffy, left-brain activity for Witches, but it is extremely valuable in helping us deepen our connection to nature. First, the process of identifying a tree or a flower will make us observe it more closely and look for characteristics we might otherwise not notice. Second, knowing the actual names of things, especially their Latin names, will help us talk about them and learn from other people's experiences. Many plants have similar common names but are actually very different.

So start collecting guidebooks and using them. The most helpful books have a key, a set of simple choices that guides you through different families and species. Does the tree have leaves or needles? Are they in bunches, or set all around the twig?

Don't get overwhelmed. There are millions of beings around us to name and identify and learn. Begin with some modest goals, such as learning the key trees, birds, mammals, and insects of your area. Or make it a goal to identify one new tree or plant on each walk. Once you know the name of a plant or animal, read up on it and learn more about it. Move from observing to understanding.

NINE WAYS OF OBSERVING

The following exercises take us through nine ways of observing. They are inspired by Bill Mollison, one of the founders of permaculture, and by the lessons I've learned from the Wilderness Awareness School. Some of them will be further developed in later chapters on the elements, but taken together, they are the beginning of learning to read a landscape.

1. *I Wonder . . .*

In your home base or other natural spot, with your attention on what is around you, say to yourself, "I wonder . . . "

"I wonder why lichen is growing on that side of the tree, only?" "I wonder why the snowdrifts are piling up in this particular pattern?" "I wonder what attracts that bug to that flower?"

Don't worry about answering your questions; just notice what questions you can generate. As much as possible, keep your questions focused on physical reality. Not "I wonder how that tree likes all that snow on its branches," but "I wonder why those branches don't break under the weight of all that snow."

This is a great exercise to use with kids. You might ask them, "How many 'I wonders' can you find in five minutes?" You could follow that exercise up at home with a session with the encyclopedia, trying to answer some of the questions. But the focus here is less on answers than on learning to generate intelligent questions.

2. *Observing Energy*
Ask yourself, "How is energy coming into this system? How is it being exchanged?" There are many different sorts of energy you might observe: sunlight, heat, energy generated by motion of air or water, food, even psychic energy (but take time to focus on the physical before you jump to the psychic.) Also, you might try sketching your spot, or a plant in it, purely as a pattern of light and shadow. Don't worry about producing a "good" drawing; just let it become a meditation on how light energy is intercepted by form.

3. *Observing Flow*
In your home base, observe flows of all kinds. How does water move through this system? How do wind and airflow affect the area? What intercepts the flows? What marks do they leave of their passage? What is the source of these flows? How is that source replenished?

4. *Observing Communities*
What is growing together with what in this area? Which trees with which bushes, which groundcovers? Are there patterns you can discern? Are there sword ferns under the redwoods, and tanoaks near the clearings? What insects, birds, and animals seem to be connected with what plants? Are some plants serving as "nurses" for the young of others? Do some plants seem to stay distant from each other? Are some plants always found together? (Note: such questions can generally be answered only by many observations over time.)

5. *Observing Patterns*
What patterns can you see there in your spot? Textures, patterns of growth, distribution patterns, stress marks, all are examples of patterns. What patterns are repeated, on what scales? Can you find spirals? Pentacles? Branching patterns? Patterns based in fours or sixes? How many times does a tree branch from twig to trunk? What functions might these patterns serve? Why are certain patterns repeated over and over again in nature?

Again, you might wish to take a session to draw patterns or forms. Put your thoughts on paper without worrying about producing a work of art, but simply as a meditation to sharpen your ability to see and focus.

6. *Observing Edges*
Where does one system meet another in your spot? As we saw earlier, edges—places where forest meets meadow, or ocean meets shore—are often the most diverse and fertile parts of an ecosystem. Is that true here? How does the edge differ from the center?

7. *Observing Limits*
What limits growth here in your spot? Shade? Lack of water? Soil fertility? Other factors? How do these limiting factors make themselves evident? What is succeeding in spite of these factors? What seems held back? How have the plants and animals adapted to these limitations? What characteristics do the successful adapters have in common?

8. *Observing from Stillness*
Just sit still in your spot for at least fifteen minutes—longer is better. Notice what you can see, and how that changes over time.

9. *Observing Past and Future*
What can you observe in this spot that can tell you about its past history, and how it might have changed over time? What can you observe that tells you something about the future of this place?

Practicing the skills of observation, taking the time to ground and listen, we begin to be able to hear something. When we clear away some of our inner obstacles so that we can open up to the outer world, when we allow ourselves to be present, we can be fed and informed and delighted by the richness of life around us.

SIX

The Circle of Life

In the Goddess tradition, all ritual takes place within a magic circle. We ground and then create a sacred space, calling the four elements of air, fire, water, and earth, and that sacred transformative spirit of the center.

The circle is the pattern of the whole, the schematic diagram that lets us know if something is complete. In *The Spiral Dance*, I discussed how to create sacred space by casting a circle, invoking the elements of air, fire, water, and earth in poetic and mythic ways.[1] In this book, now, we enter the circle, to begin a journey through the elements of life. While we know that air, fire, water, and earth are not elements in the same way that hydrogen, oxygen, nitrogen, and carbon are, they each represent great cyclical processes of transformation that sustain life. The swirling cauldron of the atmosphere, the energy exchanges fueled by the sun, the cyclical journey of water from raindrop to stream to ocean to raindrop, the endogenic cycle of rock formation and plate tectonics, and the cycles of birth, growth, decay, and regeneration are some of the most basic processes of Gaia's physiology.

Many indigenous traditions use the pattern of the sacred circle with the four elements and the four directions. Not all place the same elements in the corresponding directions, because the correspondences originate from the qualities of particular places. The correspondences I use here originate from the British Isles, where the west wind brings rain, as it does in California, where I live. The west corresponds to water, to *feeling*, to emotion, to twilight and autumn—very fitting in a land where rain returns in the fall. The north is the direction of earth and the *body*, midnight and winter. The east, the place of sunrise, is air, *mind*, thought and inspiration, dawn and spring. The south, where the sun is strongest, is fire, *energy*, high noon, and summer. The center where the elements meet is the place of *spirit*, of change and transformation, the timeless place, the heart. In another place or climate, the directions for each element might change. In Australia, for example, Witches invoke fire in the north, not the south. But regardless of such changes, the circle of the elements remains an image of wholeness, and the correspondences of air/mind, fire/energy, water/emotion, earth/body, and center/spirit hold time.

More than twenty years ago, in *The Spiral Dance*, I discussed the magical correspondences of the elements. In this book, I want to take us on a journey around the magic circle, this time experiencing the great natural cycles and life processes each element reflects.[2]

The magic circle, because it represents wholeness, is a pattern we can use for examining the whole of any subject.

The Elements of Decision-Making

When we want to know if we have considered all sides of an issue, we can think about the elements and their corresponding qualities: What do I think about this particular issue? What energy do I sense around it? What do I feel? What is my body telling me? What transformation is possible?

Or: What are the rational arguments for this proposed action? What energy will it use or generate? What flow of events, materials, or processes will affect (or be affected by) it? What are the material considerations, the constraints? What are the ethical and community concerns?

When making a decision about sustainability, for example, we can ask,

How will this proposed action affect the air, the climate? The birds and insects? Will it bring inspiration and refreshment?

How much energy will this use, and where will it come from? Will it use more energy than we take in? How much human energy will it require? Will it energize or drain us?

How will this affect the water? The fish, sea-life, and water creatures? Will it use more water than we have? How do we feel about it?

How will this affect the earth? The health of the soil? The microorganisms and soil bacteria? The plants and animals? The forests?

How does this affect our human community? Will it benefit the poorest and least advantaged among us? Does this reflect and further our deepest values? Will it feed our spirit? Will it create beneficial relationships?

WORKING THE CIRCLE

Do you have a decision to make? An issue you are considering?

Take your journal to your home base, or to a quiet, safe place. Ground and come into your senses. Now consider your issue, using one of the sets of questions above. In your journal, record your inner conversation.

Then read through the next five chapters of this book. When you're done, go back to your issue and ask the questions again, once more recording your inner dialogue in your journal.

Now compare the two entries. Has anything changed? Have *you* changed?

Casting a Circle

At times in this book, I will suggest that you cast a circle before doing an exercise. To cast a circle is to create an energetic form of protection around yourself, a boundary that can keep out interference and negative forces. Before embarking on a trance journey or a deep change in consciousness, Witches cast a circle of protection. In fact, we begin all formal rituals by casting a circle and invoking the four elements of life.

Casting the circle creates an anchor, a set of physical, verbal, and visual associations that over time facilitates your transition into and out of particular states of consciousness. You can use many different methods to cast the circle. Some good examples can be found in *The Spiral Dance* and *The Twelve Wild Swans* (which I coauthored with my friend Hilary Valentine), and I offer one below.[3] But if you find one form that you like, and use it consistently, especially when working alone, the repetition will strengthen its effectiveness as an automatic trigger to consciousness change.

FERI CASTING

Casting a circle can be a very simple procedure. You can use your hand or a magical tool—an *athame* (or Witch's knife), if you have one. I generally use my garden pruners.

Here's a very simple casting I learned from Victor Anderson, my teacher in the Feri tradition of Wicca:

First, ground yourself.

Then stand in the center of the room, facing north, and say,

"By the earth that is her body . . ."

Turning to the east:

"By the air that is her breath . . ."

To the south:

"By the fire of her bright spirit . . ."

To the west:

"And by the waters of her living womb . . ."

Turn back to the north, to complete the circle. Face center and say,

"The circle is cast, the ritual is begun. We are between the worlds. What lies between the worlds can change the world."

While you are moving and speaking, visualize a circle of light, of blue flame or whatever color of light you like best, surrounding the area you are working in.

Instead of standing in the center, you might also walk around the room and tap each of the walls in turn. If you are not a strong visualizer, physically creating the circle will make it stronger.

Once the circle is cast, try not to leave it until the ritual is over. If you need to leave, respect the circle by cutting a door into it—using your tool or hand to mime cutting an opening in the energy boundary, or taking both hands and parting the energy as if you were parting draperies. Close it behind you—and don't forget to open and close it when you come back in.

Once the form of the circle is created, invoke or acknowledge the four elements and the center, and then call in any Goddesses, Gods, ancestors, or other beings you want present in the ritual, expressing your gratitude to them for their gifts.

When the ritual is done, open the circle. You can physically walk back around the circle in reverse, or simply thank all the energies you've invoked, and each direction in reverse order, and then say goodbye. You might say,

> By the earth that is her body,
> And by the waters of her living womb . . .

By the fire of her bright spirit,
And by the air that is her breath . . .
The circle is open but unbroken.
May the peace of the Goddess go in your hearts,
Merry meet and merry part.

When the circle is cast, we are ready to meet the elements of life.

INVOKING THE ELEMENTS

There are many ways to call in the elements. You can memorize a poem or simply speak from the heart, sing, dance, drum, or whisper. But one of my favorite ways, in a group, is a small ritual Kitty Engelman and I created for a weekend workshop we taught together at Diana's Grove in Missouri.

Ask everyone in the group to be still for a moment and feel which of the four directions, plus center, they are called to. Participants then move into their specific direction and are helped to ground and come into their senses.

Everyone is then sent out to observe for fifteen minutes:

The east group is asked to observe light and motion.
The south group is asked to observe fire and energy exchanges.
The west group is asked to observe water and flow.
The north group is asked to observe earth, plants, and animals.
The center group is asked to observe patterns.

When they return, the east group is asked to step into the center of the circle and hold hands. For a moment, they acknowledge each other, then turn around and face outward. Now they are asked to tell the stories of what they observed, all speaking at the same time and speaking continuously as the circle moves around the central group.

Hearing the stories is like hearing a spoken poem, taking a journey through the images and sensations of air. When the listeners have made a complete circle, the east group is asked to turn in and take hands again, and the whole circle expresses gratitude to the air and the storytellers of the east.

Repeat that step with the south, west, north, and center groups.

Another way to do the storytelling is to have the listeners stand still and the speakers face out and slowly revolve, telling their stories over and over until they return to their original spot.

Ritual Structure

For more than twenty years, I've worked at teaching and practicing ritual with a group called Reclaiming. We began as a small collective of women and men who wanted to integrate our spirituality and our political activism, and over the decades have grown into a network that extends from California to Germany to Australia and considers itself a tradition of the Craft.

In the Reclaiming tradition of the Craft, our rituals include spontaneity and creativity, but they follow a structure that begins with grounding, cleansing, and casting a circle. Cleansing can be a meditation, a bath, a plunge into the ocean, or it can be incorporated into the grounding by taking some extra deep breaths and letting tension go. Casting the circle creates a container for the energy that will be raised. When the circle is formed, we invoke the elements of the four directions, plus the center, which is spirit, and also acknowledge and invoke any Goddesses, Gods, ancestors, or other spirits we wish to call. Then we do the work of the ritual, which can be almost infinitely varied, but generally involves some shift in consciousness or some focused direction of energy toward an intention. After energy is raised, it is grounded. Then we often bless and share food and drink, thank all the energies we invoked, and open the circle.

As we journey through the elements in the chapters that follow, I will occasionally make suggestions for rituals. Generally, I will focus on the heart of the ritual and assume that you will first ground, cleanse, cast, and call the elements and the sacred in your own way, and thank them and open the circle afterward. *The Twelve Wild Swans* also includes a full discussion of ritual structure and creation.[4]

And now let's take a journey through the elements of life.

CHAPTER 7

Air

From my journal:

I'm walking in a break in the storm. For two days it's been raining, yesterday so hard that I never went out at all. Each time I thought about a walk, the rain would pelt down in sheets. But today, as soon as it lightened up, I decided to go.

I meet my neighbor Angie halfway down the hill, and together we walk up to Transmission Hill—so called because of the sculptures and tables made of old car parts left at an abandoned campsite on the top.

At first, the long view to the north is entirely hidden in fog. But as we watch, a small window opens, like a hole in the gray through which we can see the green hills opposite us. We follow the moving hole, almost like viewing through a grand telescope. And then the fog begins to break up. First the cloud layer rises, the hills remaining shrouded while vistas open up below. In the valley, we can see the fog rolling and boiling. The vapor is so thick that the air currents are made visible, marked by shreds and tatters of gray against the dark green mountains behind. We watch them shift and dance. They look like waves, cresting and spiraling in slow motion. We can see swifter currents moving quickly down the center of the valley, leaving wisps trailing behind.

We continue on our walk, heading around the hill. Today the wind is coming from the west, bringing the rain with it. Suddenly we see a crazy raven, riding the air currents as if they were his own private roller-coaster. He, too, makes the patterns visible. He glides up and down over the great waves of the air exactly as surfers ride the waves of the sea. He soars aloft on an updraft, tucks his wings and plummets, pulling up and flipping over like a fighter plane before he nears the ground. He is having a wonderful time, showing off. Now his mate joins him, doing her own dives, tucks, and rolls. The wild storm winds have become their amusement park, their courtship ground, for this fancy flying is part of their mating dance.

Air is the most ubiquitous element. It surrounds us, and we live immersed in it. In fact, we can't live for more than a few moments without it. Invisible, it is revealed to us only through its effects on other things. We can't see the wind, but we can feel it and know the immense power of a windstorm, a hurricane, a tornado.

In traditional Craft practice, air is the element of the east, associated with thought and mind, with the rising sun, with inspiration and enlightenment. Air is the breath of the Goddess, her living inspiration.

While in these chapters we are focusing on each element individually, one cannot actually be separated from another, any more than your breath can be separated from your circulation or your bones. They are all part of the whole that is the living being of the earth.

And so this wind that carries the raven is moved by the great energetic forces at play over the sphere of the earth. The earth is like a great inside-out cauldron, with the air its bubbling, spiraling brew, heated by the sun's fire, contained not by iron walls without but by the embrace of gravity from within.

The sun pours energy out on the earth as heat and light and radiation. The tropics, the band around the earth's equator, receive that gift most strongly. There, the air is heated up, and heat rises. As the hot air moves up, it pulls in colder air from below. A band of rolling air, like an elongated fountain, stretches across the belly of the earth, creating a turbulence, a complex swirling, spiraling motion, that distributes some of that heat throughout the northern and southern hemispheres. That motion cools the tropics and warms the Arctic and the Antarctic, preventing the lands along the equator from roasting in 140°F heat, and the poles from dipping to an average of –150°F. Eventually, excess heat radiates back into space.

The winds are moved, as well, by smaller variations in energy and form. In summer, when the inland valleys heat up, warm air rises and sucks in the fog from the coast—fog that often sits like a literal wet blanket atop San Francisco. In winter, cool, moisture-laden air rides in from the coast and drops its heavy burden of rain on the slopes of the mountains as it's forced to rise.

To observe the wind, to feel it on your face or to lean into its wild power in a storm, is to literally feel the energies of the planet in motion.

WIND OBSERVATION/MEDITATION

In your home base, or wherever you happen to be, ground and come into your senses. Now focus on the air and the wind. Notice what you can smell and taste on the breeze. Where has this air come from, and what does it carry with it?

Notice the feel of the air on your skin. Is it gentle or powerful, moving or still? What is the temperature?

Listen to the sound of the wind. The wind is telling you what it's moving through, and how fast. The wind has a voice that sounds different through trees or around the corners of houses. It can tell you how much moisture it carries, and whether or not there is a storm coming. What does the wind say to you now?

Look around you and notice what is responding to the wind, and how. Are branches bending and swaying? Are there trees or bushes that show signs of having been shaped and pruned by the wind as they grew?

How does the wind respond to obstacles? Does it move differently through swaying branches than it does encountering a wall?

Get up and move around now, and find the places where the wind is strong and the places that are wind-sheltered. What makes the difference? Feel what happens when the wind hits a hard surface or a soft surface.

What can the wind teach us about movement and change? About responding to obstacles, or about softening the great forces that impact us?

Is there something in your life you'd like to let go of? Imagine that you hold it in your hand. Raise up your arm, open your fist, release it, and let the wind carry it away.

FINANCIAL DISTRICT WIND OBSERVATION/MEDITATION

Go to the downtown financial district of your city, or some other area full of high-rises and hard surfaces, on a windy day. Walk around and notice where the wind is strong and where it is blocked. Feel how the hard surfaces accelerate the strength of the wind as it's channeled through narrow corridors. Are there temperature differences? Sheltered spots? If there is a park or garden, notice if the wind changes there. Does the wind behave differently here than in your home base spot? How does sound travel here? What smells does the wind carry?

Now sit in a safe place and close your eyes for a moment. Think about what you have observed. Are there parts of yourself that, like the wind, get blocked in this environment? Are there ways in which they gain strength or force in response to the blockage?

The wind is wild. Even the most urban environment cannot keep out the wind. Breathe in some of that wildness, and let it feed and strengthen the wild parts of yourself.

Microclimates

From my journal:

I stand on a ridge, a thicket of mature tanoaks standing between me and the wind. I can see their branches waving, undulating, tossed by the wind, but I feel only a cool, gentle movement of air, not so much a breeze as a quickening. The trees filter the wind. The shapes they make through space, the thousands of leaves and twisting twigs and branches, intercept the wind's energy and slow it down, trapping it in a labyrinth.

This windbreak has such a different quality from the way a solid surface interacts with the air—the difference between a seduction and a slap in the face. I'm thinking of the way the wind is funneled in San Francisco's downtown financial district or the middle of Manhattan, between tall buildings, hitting against hard edges, and how it feels to step around the corner of a building into a sharp-edged gale. The trees, with their filtering effect, slow the wind down. The sharp-edged buildings speed it up.

I continue on my walk, noticing how the rounded shapes of the mountains create wind shelters that are subtle. From the top of the ridge, I can look out for miles and miles. The air above is a layer of gray, the cloud-forms in ridges and ripples. It's like looking up from undersea and seeing the waves from below. Moisture-laden air is water, moves like water, only it's not as dense.

Walking around the bulk of a round hill, I'm wondering which direction the wind is coming from. Not from the west today. At first it seems to be coming from the east, or even up from the south; then it slacks off as the bulk of the land shelters me. Now I round the corner and suddenly the wind is in my face, coming down from the north, and I can feel how the hill has split the wind into two streams as a boulder in a river divides the current. The wind is more forceful here. This hill is bare, with low grass and few trees, logged and then grazed for a hundred years. And this south-facing slope is oak savanna, not forest. Huge black oaks and valley oaks, bare now of leaves, reveal the direction of the prevailing wind in the patterns of their branches. The wind has sculpted them into a form which feels like motion even when the air is still and which speaks of the west wind even when, as today, the wind is blowing from a different direction.

The removal of trees has changed the way air moves over this landscape. The wind is faster, harsher, more erosive. I think about the scale of deforestation happening all over the world today. How does it change the wind patterns to remove millions of acres of trees? With the buffers gone, the winds speed up, and the great worldwide currents of air may themselves be changed. No wonder the weather is always unusual these days!

To understand how air moves, picture molasses or honey oozing over an uneven surface. Cold air will flow downhill, as molasses would, but ridges or bands of trees can trap it, creating cold pools. Warm air will flow upward, warming the ridges, creating updrafts that hawks and vultures ride aloft. These small variations in temperature and wind strength create varied living and growing conditions.

A backyard, a city street, a forest is full of these areas of small differences, called "microclimates." One area of your yard might be perfect for sunflowers, while a few feet away you might be able to grow only ferns. A north-facing city park in cool, foggy San Francisco may get very little use compared to its sunny, south-facing neighbor. The daffodils on my sunny ridge may bloom a week sooner than the ones planted down in the shady canyon.

Microclimates are useful when we think about how to put things in the right place. If you are planting an apricot tree, for example—one that tends to bloom out early in response to the first spring sun and then get caught by frost—you might want to root it in a cold hollow where it won't warm up so early and thus will delay its blossoming until warm weather has firmly arrived. If you are building a house or planting a garden, you will want to carefully consider the microclimate of your location.

One cold, north-facing park in downtown San Francisco gets no lunchtime use because it's never warm. It's marked by a black stone sculpture, popularly called "The Banker's Heart," and it feels and looks cold. In our foggy city, people seek out the sun. In the blazing heat of a Texas summer day, on the other hand, shade would be an attraction.

If you are planning a ritual, you probably want to find a spot that is wind-sheltered but warm. That enticing valley in the park could be a cold sink on a May morning. That sunny hillside may be swept with winds that will tie your Maypole ribbons into Gordian knots. Consider where the sun will be at the time of day you'll be gathering. That shady glen might turn into a blistering inferno in the afternoon. If possible, always scout a location at the time of day you intend to use it before you finalize your plans.

MICROCLIMATE OBSERVATION

You can conduct a microclimate observation in your home base or in the city, but it is most interesting in a hilly area with different elevations or patches of sun and shade. Walk around the area and notice the changes in climate and temperature with changes in elevation, amount of shelter from the wind, amount of shade and sunlight. Are these changes reflected in different vegetation? Different ways humans use the area?

Where do you feel most comfortable? At ease? Where do you feel invigorated? Where would you want to be on a hot day? A winter lunch hour?

Consider some other factors of this microclimate. The same factors that amplify wind can amplify noise. A wind-sheltered area may also be a spot where smog collects.

How does the microclimate affect the more subtle energies that you feel? If you have a chance to visit a sacred site, what can you observe about its microclimate and relationship to the land?

Shelter from the Storm

From my journal:

The wind blew down the yurt we built on a sunny hillside to house friends and caretakers who could watch our garden when we were gone. A yurt is a canvas structure originally invented by the Mongolians as a nomadic shelter. Its walls are a lattice—like an expanded version of an oversized accordion-style baby gate. A cable sits atop the lattice, and the rafters slot into the cable at their tails and tuck their noses into a central ring, which in the modern version is topped with an acrylic plastic skylight. Tension and compression work in balance to keep the thing from falling down. It all gets covered with a skin and tied down firmly. In theory, it's a very wind-resistant design—and certainly the Mongolians have plenty of wind to contend with.

We had gotten as far as the "put the thing up" part (the work of many hours)—but not as far as the "tie down firmly" part. We were then gone overnight one day—a night in which fifty-mile-an-hour winds howled over the hilltop. Our yurt platform was nestled into the side of the hill—a location which, we had hoped in the planning, would give it some shelter from high winds. Not enough, evidently. The neighbors reported seeing it billowing in the wind. We returned to find it smashed flat, the skylight cracked, the roof ripped to pieces, a few rafters broken, and many pieces of lattice cracked.

A sad and discouraging climax to months of work. The deck held firm, however. The months of work had been mainly devoted to constructing the deck platform, which our neighbor Dave had designed for us with an extended concrete foundation to provide strength against the wind. Hundred-mile-an-hour winds have been known to rip across that hillside—one actually blew off the roof of a small cabin when Doug, who built it, first lived there.

I'm now amazed at our lack of clear thinking—putting up such a strong deck and not considering the strength of the yurt itself. We were lulled into a sense of false security by the yurt brochures, which advertise the structures as being designed for wind resistance. True—when they are fully tied down. Nothing was said about the wind resistance—or lack thereof—of a half-finished yurt.

So—we've ordered new parts, with the help and sympathy of the folks at Pacific Yurts, who scoured their warehouses for seconds we could buy at a discount. The yurt will rise again!

I hope most of you will never learn quite so dramatic a lesson about the power of the wind. But anyone who has ever keeled over in a sailboat, attempted to walk or cycle into the teeth of a strong wind, or been buffeted on an airplane knows how strong and sometimes terrifying the wind can be.

Our society expends a huge amount of resources and energy keeping buildings cool or warming them up. Much of this expenditure could be eliminated if we understood better how to create shelter from the wind.

Although our yurt lasted only a day before succumbing to the wind, we have a greenhouse nearby—a sixty-foot-long womb-shaped bubble of thin plastic stretched over metal rods (much lighter than the unfortunate yurt!)—and it has withstood many fierce storms. Why so hardy? It sits behind a sheltering belt of trees. I chose its location because I always noticed, when walking along that stretch of road, how the wind died away as soon as I moved behind the trees.

Understanding how to create shelter from the wind can help us understand better how to withstand any strong force, physical or emotional.

Our usual response to a force is to try to block it. To shelter a patio, we build a wall. To withstand a criticism or an attack, we go cold, shut down, turn to stone, or otherwise defend against it.

You may have noticed in your downtown observations how the hard surfaces and high walls don't so much *stop* the wind as *redirect* it. Wind is embodied energy—when it hits a hard surface it bounces off just like a basketball hitting a backboard, and it heads in a new direction. Impervious walls create eddies and new turbulences that may be just as harsh as the original ones you were trying to block, and more unpredictable.

The elements are our teachers, in many ways. Consider for a moment whether you've ever reacted defensively against a strong force in your personal relationships. Have you built a wall to keep out someone or some group you considered different or threatening? Have you tried to shut out a criticism or block an attack? What happened?

Trees don't place an impervious barrier before the wind. Instead, they absorb and transform the wind's energy. Watch the branches sway. Each one is like a springboard, set in motion by the wind, using the wind's energy to move itself in a new dance. First the branches go with the wind, bending in its direction. Because wind is moving in spirals and eddies, the branches find an ebb moment to spring back again, only to be pushed yet again by a new wave. That reverberation uses energy, and if there are enough branches, the wind's energy will be used up, or at least diminished, its force softened.

A windbreak is a useful thing to know how to construct. If I want to shelter my home, my ritual site, my yurt from the wind, I won't build a wall around it; I'll plant trees and bushes, or set up a filter with moving parts to absorb the wind's power. By doing so, I can make my home or garden more comfortable, more energy-efficient, and safer. I'll need to expend a lot less on heating and cooling and replacing broken windows.

But a windbreak is also a great teacher, for me, about nonviolence. How do we respond to strong forces—anger, rage, even physical attack—without becoming violent in return? How do we respond to what might be well-meant but harsh criticism (whether well intentioned or intentionally hurtful)?

If we become a wall, shutting out the energies coming at us, we may actually strengthen the anger of the opposition. On the other hand, if we simply brush off or bat away criticism, the opposition may expand its criticism to include our reactions.

But there's a third alternative: if we can learn from the trees, we can take in and transform the energy coming at us. We do this by staying calm and grounded and centered, by listening rather than responding, by swaying with the wind and letting it blow itself out.

The exercise below is similar to one I often use in nonviolence trainings. In including it, I don't want to suggest that this is the *right* way or the *only* way that we should respond to attacks. It is simply one of our options, and the purpose of learning it is to expand our range of choices in any situation.

WINDBREAK/NONVIOLENCE EXERCISE

Do this role-playing exercise with a partner, or if you're in a group, have people form two lines, facing each other, with partners across from each other. The ground rules are simple: interact only with your partner, stay with the interaction and don't walk away, and don't use any physical violence in the exercise.

Choose two oppositional roles that relate to a situation you might face. In preparing for political actions, for example, one side might be peace demonstrators; the other, angry supporters of an invasion of Iraq. But you can also choose everyday situations. One partner might be a dogwalker; the other, an angry homeowner accusing the walker of letting the dog foul the sidewalk. One partner might be a leader of a public ritual; the other, a police officer who's gotten a call complaining that Satanic rituals are happening. You can also use this exercise to prepare for an anticipated difficult encounter: one partner might be a newly enthusiastic Witch going home for Thanksgiving dinner; the other, her belligerent fundamentalist brother-in-law.

The partner who is the windbreak should ground and center. The attacking partner should take a moment and think about the aggressor's role. How would you feel in that role? Have you ever been angry or wanted to attack someone? What did you do? Say? What kind of energy did you feel?

At a signal, begin. Let the interaction go for two to five minutes, with the attacking partner putting into words and gestures the anger of the role, and the windbreak partner focusing on listening and on staying grounded and centered. The windbreak partner may speak or keep silent.

Then stop the interaction and debrief. Ask the attacking partner,

How did you feel?
What did your partner do? Did she or he do anything that was effective in communicating or deescalating your anger?
Did she or he do anything ineffective? Anything that made you angrier or cut off communication?

Ask the windbreak partner,

How did you feel?
What did you do that felt effective? Ineffective?

Were you able to stay grounded and centered?
What did it feel like to absorb the energy of your partner's attack?

If in the course of the exercise you took on any negative energy that wasn't yours, take a breath and consciously release it, or do an energetic cleansing. (See the exercise called "Energy Brushdown" in Chapter Eight.)

Air and Life

Air is life. Not only does most life on the planet need air, but the atmosphere itself is a creation of life. Remember the creation story we read in Chapter Four? To understand the miracle gift of the atmosphere and the incredible dance of life that created the very air we breathe, do the following meditation.

BREATH MEDITATION

Read the story of "The Gift of the Ancestors" or have someone speak it as a guided meditation. Now close your eyes for a moment and simply breathe.[1]

Breathe in with gratitude to the ancestors and with appreciation for the great creative powers of life.

Breathe out with love, as a conscious gift back to the green world.

This air is a gift of the early ancestors. What you breathe in, this moment, originated billions of years ago. This air passed through the lungs of dinosaurs and mammoths and the earliest human beings. Is there a great teacher or hera from the past that you admire? You are breathing the same air that passed through her or his lungs, sharing inspiration. Breathe in with gratitude; breathe out with love.

Is there a problem you're stuck on, an issue that seems hopeless? A place where you've given up in despair?

Just breathe in, taking in the creative power of the ancestors, of life, asking that power to infuse you and help transform your issue.

Breathe out, with love and commitment, acknowledging your grief or pain or hopelessness, not trying to change it but just making space within, to breathe in again, filling yourself with creativity and power.

Continue as long as it feels right. Don't try to solve your problem instantly; just keep breathing into it and trust that you can shift the energies around it and open the space for inspiration to come.

The Birds and the Bees: Creatures of Air

The air not only sustains our life; it is a medium through which life moves. The birds and insects can be our teachers, too.

Philosopher and student of shamanism David Abram, in *The Spell of the Sensuous*, writes of a mystical experience he had in Bali one night, watching the brilliant stars of the night sky reflected in the still waters of the rice paddies below, feeling as if he were free-falling through space:

> I might have been able to reorient myself, to regain some sense of ground and gravity, were it not for a fact that confounded my senses entirely: between the constellations below and the constellations above drifted countless fireflies, their lights flickering like the stars, some drifting up to join the clusters of stars overhead, others, like graceful meteors, slipping down to join the constellations underfoot, and all of these paths of light upward and downward were mirrored as well in the still surface of the paddies. I felt myself at times falling through space, at other moments floating and drifting. I simply could not dispel the profound vertigo and giddiness; the paths of the fireflies, and their reflections in the water's surface, held me in a sustained trance. . . .
>
> Fireflies! It was in Indonesia, you see, that I was first introduced to the world of insects, and there I learned of the great influence that insects, such diminutive creatures, could have on the human senses.[2]

Birds are easy to appreciate and love, but insects are a challenge for many people. They seem strange and creepy. Science fiction aliens and villains are often portrayed as insectlike. In the real world, some insects sting, bite, or nibble on us and our belongings in unpleasant ways. Some are poisonous and others carry diseases.

I know too many Witches who rarely go outside because they are afraid of bugs. How sad!

We can't connect with nature without connecting with bugs. After all, about half the living things in the world are insects. Most of the plant life that we love coevolved with insects. The flowers, the fruits, and the insects that pollinate them, and the birds and animals that eat the insects, make up a whole that cannot be divided.

If insects frighten or horrify you, perhaps it is because you are being a wall instead of a tree in relationship to them. Are you shrinking away from them, trying to wall yourself off from any encounter?

Try letting insects into your awareness in gentle, nonthreatening ways first. Read about them. Pick one insect to start with, and learn about it. Get to know the life cycle of a butterfly or the strange marital dynamics of a praying mantis. When you're ready, try this observation exercise.

INSECT OBSERVATION

Sit in a garden (preferably an organic garden) or other natural spot and watch the insects. How many different insects can you count in a five-minute period? The number of insect species will tell you something about the amount of life energy in this spot.

What are the bugs doing? Are they feeding? If so, on what? Are different insects feeding on different plants? Are they drinking the nectar of flowers? Are there honeybees present? If so, are they collecting pollen? (You'll be able to see the heavy pollen collected on their legs.) What plants do they seem to prefer?

Go back to this spot in different seasons, and at different times of day, and notice how the insects change.

David Abram also goes on to suggest that shamans work not so much by contacting supernatural forces as by working with the powers and awareness of the insects, birds, animals, and other nonhuman beings around us. Insects know things that we don't and perceive things that we can't. Once we develop friends and allies in the insect world, we can experiment with augmenting our awareness.

CHOOSING AN INSECT ALLY

When you are comfortable with the insect observation, choose one insect to study. Learn about its habits and life cycle. Insects undergo amazing transformations as they develop through their life cycles. Bees, ants, and termites live in complex societies, communicating with each other through scent and the amazing symbolic communication of the bee dance.

Find a spot where you can observe your chosen insect. Ground, center, and come into your senses. Watch your insect for a bit; then close your eyes. Imagine you are face-to-face with your insect and can speak to it. Ask its permission to borrow its awareness for a moment. If you sense that the answer is yes, breathe deep and imagine your insect growing larger and larger, until you can step inside it. Feel your skin transform, grow wings and antennae; let yourself see with its eyes, hear and smell and taste as it does.

At first, you may feel as if you are simply making up an experience, telling yourself a story. Don't worry. Just enjoy the experience, noticing what's different in this mode of perception without worrying about whether or not it's "real."

Begin with just a short time, and later expand it if you feel comfortable. Before you come back to yourself, thank your insect. Imagine yourself stepping out of its body, and let it shrink back to normal size. Pat your own human body; feel your skin, your human form. Use your anchor to your grounded state; open your eyes and say your human name.

When you become comfortable and familiar with this exercise, experiment with using it to get information. You might explore your garden as a bee, for example, to see whether you've provided good habitat for pollinators. I admit that I find being a bee extremely erotic. Imagine smelling something wonderful, sipping on something sweet, healthy, and nourishing, while being stroked and caressed by velvety, feathery stamens. For that matter, each of those flickering fireflies is calling out some insect version of "Hey, hey, hey, baby—I'm the guy who can show you a *real* good time!" And then there are all those damselflies flitting about, stuck together, flaunting their relationships in an utterly shameless fashion.

A little of this exercise, and you may find your horror of bugs changing to a very different emotion.

The Language of Birds

Our prince, in the story told in the first chapter of this book, takes many years to learn the language of birds. If you hope to master that language, perhaps you'd better begin right now.

The teachers at the Wilderness Awareness School say that when songbirds sing, they are giving praise to the Creator. The Mohawk people teach that the songbirds were each set in a particular place on the earth, to sing about that place and teach us something about it. That may be, but it is my personal belief that the blue jays are here to teach us that no matter how beautiful everything is, someone will always find something to complain about.

Paying attention to the birds will tell you many things about what's going on in the world, and about your own state of being. Birdsong changes according to the time of day, the state of the weather, the season of the year, and how the birds happen to be feeling. The Wilderness Awareness School identifies five key voices of the songbirds: the call, the song, the feeding plea, male-to-male aggression, and the alarm call. A real expert in bird language can tell by the birds' vocalizations what predators are moving through the forest at any moment. Scouts among the Apache could tell when a European was three miles away by the song and behavior of the birds.

Learning the language of the birds is a long and complex process, as our fairy tale suggests. If you are truly interested, I strongly suggest that you enroll in a program like that offered by the Wilderness Awareness School. (See the section on Resources in the back of the book.)

But even without that formal training, all of us can begin to hear something.

LISTENING TO THE BIRDS

In your home base, ground and come into your senses. Now focus on your sense of hearing. What birds do you hear? How many different songs or calls can you hear? Do you know what species they belong to? If not, just give them names of your own, as I did in the observation that begins Chapter Five (the Eight-Note Bird and Wheet).

Following are some suggestions for deepening your awareness of the birds. You may want to follow some or all of them.

Where are the birds singing from? Draw a rough map of your home base, and mark approximately where you hear each bird.

For a week, come back and listen, referring each time to your map. Do you often hear the same bird in the same place? Can you identify a home base for that individual, a possible nesting place?

Keep a bird log of all the different birds you hear. Again, if you don't know the species, give it a name of your own. Try logging birds on one of the following schedules:

For half an hour at the same time of day each week (and again in different seasons)

For half an hour at dawn, just after sunrise, and at midday, twilight, and darkness

On the first day of each month, over several years (to learn if the species count changes over the years)

Few of us will be quite so thorough in our observations, but just thinking about these questions will help us sharpen our awareness and take in more information.

The Five Voices

Learning to recognize the five voices of the songbirds—the call, the song, the feeding plea, male-to-male aggression, and the alarm call—takes nothing more than time, practice, and a bit of empathy. Remember that these five voices apply only to the songbirds (the passerine, or perching birds)—not to the corvids (the crows, ravens, etc.) or the jays, who are a law unto themselves, nor to the raptors, the birds of prey.

Now as you listen to the birds in your home base, you'll be able to identify the following voices.

The Call. The call is the bird's basic "Here I am" statement. It's regular, repeated, and often echoed by a mate, as if they were saying, "I'm here, sweetie; everything's fine." "I'm here, too; all's well." "I'm still here; everything's fine." "I'm here, too; all's well."

The Song. The song is generally more elaborate, sung especially in the spring. Many birds burst into song with the sun's rising, and why not believe that they are filled with joy and delight and gratitude for a new day, singing their general well-being and thanks? Western scientists, however, maintain that they are defending their turf and advertising for mates, that their song is the bird equivalent of "Single male robin with good breeding territory seeks committed relationship. If you thrill to a sunrise and enjoy a good, juicy, early worm, if you dream of sharing a nest and rearing a clutch of eggs . . . let's meet for a twilight flight and explore possibilities."

The Feeding Plea. You're likely to hear this mostly in spring and early summer. It's the begging cry of baby birds, generally high-pitched, frantic, and recognizable to any parent: "Hey Mom, I'm hungry! Dad, more worms! Where are you guys? I'm starving! Feed me! Feed me!"

Male-to-Male Aggression. This too is a spring/early summer mating and nesting thing. It can sound like an alarm, but it's limited to one species of bird. Again, it's not hard to recognize or identify with: "This is my turf!" "Yeah, well I'm taking over; move on out!" "You move, you ***!!!*%%**." "Take that, you !!***!!*%%**."

The Alarm Call. A true alarm spreads from one species to another and often moves out over the landscape, following the path of a predator. It might be a bird's normal call, speeded up or more frantic in pitch. Squirrels and jays often join in. Again, a real expert can tell exactly what is moving through the forest by the nature of the alarm calls. Personally, I'm still at the stage of "Hmmm, something is sure upsetting that chickadee. I wonder what it is." An alarm sounds like, "Look out! Look out!" "Head's up, everybody!" "Watch out! Watch out!"

An alarm in the forest tells you that some predator is on the prowl. It might be dangerous to you: a cougar. It might be a danger only to the birds: a hawk or a housecat. It might be something simply out of place: another human in a bad temper.

What the Birds Are Saying About You

Yes, the birds are talking about you. Whenever you go into the woods or out to a meadow, the birds notice you and they respond to the state of consciousness you're in. If you are crashing through the woods, talking loudly, or *thinking* loudly about how mad you are at that person in your office who sent you the nasty email, you probably won't see or hear many birds. They will simply flee, and you will walk through a silent landscape, probably not noticing the birds at all.

When you are grounded, in your senses, practicing your awareness techniques, and attempting to walk silently and respectfully, the birds will let you come closer. Eventually, they'll simply "hook"—that is, move to a minimally safe distance away. When you gain enough practice, awareness, stillness, respect, and love, they may even approach or not change their behavior at all.

The birds and animals will be our teachers if we let them. The other day I was walking in the forest, stepping as silently as I could on a bed of fallen leaves, practicing being in my senses, not my thoughts. All was going well until my mind strayed to the upcoming WTO protest in Cancun, and the likelihood of police violence there. At that moment, a squirrel leaped onto a branch above my head and began scolding me, deeply affronted. It was as if he were saying, "How dare you bring those nasty thoughts into the woods! Here you were, walking so quietly and we thought you were a nice, safe human, and then you come out with *that!* I'm shocked at you! And just when we're rearing young ones!"

I stood still and apologized, but it did no good. He continued to follow and scold me until I was out of his home territory.

On the other hand, I was on the same trail one day, passing through the same rustling dry leaves, when my mind strayed to the possible scenario for an erotic film my friend Donna and I are always threatening to make. A whole covey of quail came out of the tanoaks and continued feeding very calmly just a few yards away from me, paying me no mind whatsoever.

CONVERSATIONAL EXCHANGE

As you sit in your home base or walk through the woods, notice how the birds and animals are responding to you. Do their responses change as your awareness changes? Your emotions? The content of your thoughts? As your ability to ground and be in your senses deepens, how do the wild animals mirror that change?

Global Warming and Climate Change

So here we are, down at the bottom of this swirling cauldron, utterly dependent for our very lives on the constant re-creation by life itself of just the right amount of oxygen in the air. And what are we doing? We are daily pumping toxic chemicals into the air, filling it full of substances that destroy the ozone layer, the protective membrane in the outer atmosphere that shields us from the most powerful and dangerous radiation from the sun. We're burning the remains of ancient organic life at such a phenomenal rate that we have increased "greenhouse gases"—gases that prevent radiation from escaping back into space—and are threatening the overall balance of the earth's climate.

"Global warming" is perhaps a misleading term, because it leads us to benign fantasies of sunbathing on otherwise chilly winter beaches and growing mangoes in Kansas City. But what global warming really means is turning up the heat on the cauldron. If you build up the flames beneath a cauldron of soup, the liquid inside boils faster. If the flames are too high, the soup may spill over the sides of the pot. In terms of the earth's atmosphere, the increased energy from global warming means that all of the turbulence of the winds is moving faster, with wilder oscillations and greater extremes. In some places, that might mean warmer winters. In others, it might mean hurricanes. In still others, drought. The patterns of climate we have planned for and adapted to can no longer be counted on, and the new patterns are likely to be more fierce and more extreme.

The earth as an organism can certainly survive this overheating, but it's not clear yet that we as a species will survive—at least not with the level of comfort and abundance we would like to enjoy. And many of the living communities of the earth that we love will be threatened, in part because so many of them are already suffering from loss of habitat, diversity, and resilience. Natural communities are capable of changing—indeed, have always changed and evolved—but they are not necessarily able to adapt as quickly as we create potential disasters. And many of the mechanisms of adaptation are blocked by human activity. Animals cannot easily migrate north or south, for example, and forests cannot retreat to higher latitudes, when freeways and housing developments and cities block their way. In earth's history, there have been several periods of great extinctions, when major life communities failed to adapt to sudden changes. We may be headed for, or already in the midst of, another such period.

We now know that global warming and climate change are underway. No impartial scientist doubts the data any longer. We see their effects in changes in the weather, in floods and hurricanes and increased storm damage. But so far the impact has not seriously affected most of us who live in the country with the greatest responsibility and lack of accountability for the problem: the United States.

But the effects of global warming and climate change won't necessarily remain as mild as they have been thus far. Ecological change is not always a gradual affair. It's more like a river heading for the rapids. Until the moment you go over the falls, the water may seem slow and peaceful, the current relatively mild. Your only warning may be the dim roaring in your ears—easily drowned out by the CD player you've brought along in your canoe. When you look ahead, the rocks and rapids are below your line of sight, and everything looks calm. And then suddenly you're over the edge and there's no turning back.

What might some of those damaging effects be? To put this in perspective, it is estimated that the earth's climate may increase by two to ten degrees on average over the next fifty to one hundred years. (A five-degree *decrease* was enough to cover the major landmasses of the earth with ice during the last Ice Age.) If ocean levels rise along with temperatures (a likely consequence, given the melting of ice), flooding will become endemic and many of our major cities will be threatened. Don't think that putting sandbags along the seafront will save us. When ocean levels rise, rivers cannot drain as easily, so inland flooding increases. We've seen that already in Bangladesh and other areas of the world.

Some places may find themselves warmer, but others may suffer a reverse effect. The Gulf Stream warms Ireland, Britain, and western Europe. Climate change could divert it, leaving the Emerald Isle with the climate of Siberia.

It is a sad failure of our current political system and our international diplomacy that we seem incapable of facing this issue. As long as the river seems placid, it is all too easy to ignore the voices warning of the rocks below, especially when what they recommend might be inconvenient or threaten the immediate profit balance of major corporations. The Kyoto treaty to limit the production of greenhouse gases and forestall climate change, itself merely a Band-Aid for the issue, was first shafted by the maneuvering of the Clinton administration and then dealt a fatal blow by Bush's outright refusal to sign.

You don't have to be a black-flag-waving anarchist to be outraged by this shortsightedness. Anyone who loves capitalism should be especially maddened—because solutions and alternative sources of energy *do* exist that could enable us to transition swiftly from our fossil-fuel-based economy to one that runs on clean, renewable energy sources that don't contribute to global warming.

There are things that we can do. First, if this whole discussion is making you feel angry, afraid, frustrated, and hopeless, take a deep breath and release some of that energy. Go back to the earlier breath meditation and breathe in some inspiration from those amazingly creative simple cells that are our ancestors. If they could figure out photosynthesis, a process so complex that scientists can barely describe it, we can figure a way out of this crisis, too.

We can make personal choices that reduce the greenhouse gases we each produce. It's important that we do this not out of a sense of guilt or resentful obligation, but as an affirmative choice to more deeply integrate our values and our everyday actions. Sometimes it's hard to believe that riding the bus on a given morning instead of driving will make a difference. But it will—especially if we make that bus ride a spell, an enacted prayer, for balance.

A SPELL FOR BALANCE

When you are faced with a choice to do something small in service of the earth that requires some sacrifice on your part, stop first, breathe and ground, and say:

"I offer up this [walk, bus ride, extra dollars spent on a compact fluorescent bulb] to the greater balance and healing of the earth. Let this small act be like a ripple that grows into a wide circle as it moves outward, leading to greater change. With every breath in, as I [walk, ride, enjoy the light], I will remember my gratitude to the ancestors for the miracle gift of air. With every breath out, I will renew my love and commitment to the balance."

The personal choices that we all make are deeply important, but change is also needed on a larger scale. So consider what groups or larger communities you belong to that might be able to make collective choices that favor the earth. Can you get your workplace or school to begin conserving energy? Switch to compact fluorescents? Run its diesel buses on biodiesel? Can you advocate for more public transportation in your community? Help educate people on this issue?

Although the Bush administration has refused to sign the Kyoto treaty, many individual cities and counties have agreed to reduce consumption to the levels mandated by Kyoto. A group known as Cities for Climate Protection is coordinating these efforts.[3]

Organizing, going to meetings, doing all the unglamorous work of political change may not seem as inspiring as ritual or a wild meditation on the wilderness winds. But these tasks can be deeply spiritual acts of service that help bring us into alignment with our deepest values. Try the blessing above before making a phone call to your representative, or in the middle of a trying meeting.

Weather-Working

As you can imagine, if you've come this far, I'm not going to suggest magical techniques for controlling the weather. The great energies that move the storms and the wind are beyond our control. Long ago, one of my first teachers in the Craft said to me, "If you want to work the weather, make friends with the clouds." If you have practiced the observation techniques in this chapter, if you have been offering blessings and gratitude to the air and breathing out your love, if the birds and animals are no longer fleeing at your approach, chances are that the clouds are feeling friendly to you. Or to put it another way, you and the air and the clouds and the creatures are now a more harmonious whole, a whole that includes the possibility that the weather will to some extent reflect your state of awareness. At my friend Penny's fiftieth birthday party, the sun shone brilliantly even though her garden near the coast is often foggy. "Who would expect anything else for Penny's birthday?" one guest remarked.

I don't like to work the weather—it seems like hubris to disturb such huge forces for my own ends. But I often find that the weather works itself—as during our ritual near the Irish ring fort when the sun came out (mentioned in Chapter Two). And when there is a true need, I simply put in a request. I stop, take a breath in with gratitude, breathe out as a conscious gift, and allow my mind to contemplate the awareness of the wind and clouds, to acknowledge that those great forces also partake of consciousness, allying with it, feeling love and appreciation for it, saying something like, "Beautiful clouds, wild wind, I love your strength and energy, and the rain that you bring is life itself. But if it doesn't disturb the greater balance, it sure would be nice to have a bit of sun and no snow tomorrow for the big peace march."

And, as friends do, the clouds generally cooperate when they can.

BLESSING FOR AIR

Praise and gratitude to the air, the breath of the living earth. We give thanks to you for our lives, for our breath, for the literal inspiration that keeps us alive. Praise and gratitude for those ancient ancestors, the first magicians, that learned to use sunlight to make food, and so gave us the gift of oxygen. Praise and gratitude to those who learned to burn food for energy, and to the great exchange, the world breath that passes from the green lung to the red and back again. Praise for the sun that sets the cauldron of the winds in motion, and to the great winds that soar over the face of the earth. Praise to the storm that brings the rains, the water of life to the land. Gratitude to the creatures of the air, the birds that lift up our hearts with their songs, the insects in their erotic caress of the flowers—a caress that brings the fruit and the seed.

May our minds be as clear and open as the air; may we learn from the wild winds how to soar across barriers and sweep away obstacles. May the air and the winds of the world be cleansed. May we learn to be good guardians and friends and allies of the air that is our life; may we make the right decisions that can restore the balance. Blessed be the air.

EIGHT

Fire

From my journal:

I'm sitting in my home base, on my back deck. Around me are clumps of redwood trees, young ones regrowing in circles around the bases of old stumps logged long ago. Above my head stretches a tanoak, and across the garden and the dry stream are beautiful big-leaf maples. Flies buzz around, and in the distance a western flycatcher calls, "Too-wheet! Too-wheet!" A raven croaks high overhead, and I hear the whuff, whuff, whuff *of its wings beating through the air. I'm thinking about energy.*

The trees all around me are capturing the sun's energy. I can see how they fill space with their leaves and needles, taking up every possible plane of interception of a photon of sunlight. That light is the power, the energy, that lets them take carbon dioxide from the air and water and make wood. Something solid out of liquid, gas, and ephemeral light.

The forest, the garden area below me, are a mass of green. All around me are organisms using sunlight to make food. And buzzing around me and crawling below me are the organisms that eat that food, and the eaters of those eaters, each a stage in the great transformation.

I have been a photon of light.

I have been a green leaf on a tree.

I have been a caterpillar munching on a leaf.
I have been a songbird dining on a worm.
I have been a hawk eating a bird.

I have been a photon of light.
I have been a blade of grass.
I have been a deer grazing in a meadow.
I have been a cougar, culling the herd.

I hear the beat of the raven's wings. It has a steady rhythm, a certain pace that I recognize as the baseline of the raven. As a drummer, I could reproduce it, hold it steady. If I had a watch, I could time it. That baseline pace represents a certain energy relationship—the size and wind resistance of the raven in relationship to the efficiency of its heart and bloodstream and the conversion of its food to energy. That pace doesn't vary much from raven to raven—not as much as, say, my comfortable rate of walking varies from that of my lithe and faster friends. But then, maybe we're just at different ends of a human baseline, which may seem to vary a lot but doesn't really—not if you compare our walking gait to that of a wolf trotting or a snail inching forward. Suddenly I'm remembering a joke told to me by a Swiss friend from Bern, whose inhabitants are said to be very slow. A Berner arrives home from Zurich, looking very exhausted and out of breath. "What happened?" asks his friend. "Why do you look so tired?" "Oh, it was terrible," the Berner says. "A snail was pursuing me all the way home!"

I'm thinking about how trees seek the light. In my city garden, our big plum tree fell down this year, on May Eve. For years I'd kept it pruned back, trying to keep some sun for the rest of the garden, but each year it grew and grew, extending its branches out to the south and west, reaching for light. All this winter, I'd looked at it and planned what branches to cut, but each time I intended to get to it, I ended up organizing some peace demonstration instead, trying to prevent a different kind of fire. So on May Eve, with the ground soft from late rains, it simply gently leaned over, pulling its roots out of the ground. We probably could have saved it even then, but I was away, and when I had time weeks later to deal with it, the roots were already dry. And the space and light that opened up in the garden with it gone were, we admitted, attractive.

So now the tree has been cut up and most of it bundled away. Instead of having plums this summer, we will have wood all winter, to burn in the woodstove, returning the tree to the heat and light, the energy, which created the wood in the first place.

Energy is fire.

The Earth's Energy Flow

With a few very obscure exceptions, like thermophilic bacteria that live in hot vents of volcanoes and such, all of Gaia's life eats sunlight—or, more accurately, lives from the food created by green plants from carbon dioxide and water, using the energy of the sun. Some forms of life, the green plants and algae, use that energy directly to sustain their lives and feed their growth. Other things feed on the plants, and still others on the plant-eaters or even other animal-eaters. But in four or five steps at most, we go back to the sun. All of life is, in a sense, a transformation of energy into form. We are sunlight at a vast costume party, dressing itself up in one form after another, discarding one outfit to pick up a different one.

That sunlight is the great grace, the one free gift that allows not just sustainability, but abundance, growth, increase, more-than-enoughness. It falls on the planet daily, without our having to do anything about it, and billions of green plants are hard at work converting that energy into various other usable forms. So don't say no one ever did anything for you!

In traditional cultures, the shaman, priest/ess, Witch, or healer was responsible for making sure that the human community remained in an energetic balance with the environment around it. David Abram, in *The Spell of the Sensuous*, writes about the shamans he witnessed in Bali and Nepal:

> The traditional or tribal shaman, I came to discern, acts as an intermediary between the human community and the larger ecological field, ensuring that there is an appropriate flow of nourishment, not just from the landscape to the human inhabitants, but from the human community back to the local earth. By his constant rituals, trances, ecstasies, and "journeys," he ensures that the relation between human society and the larger society of beings is balanced and reciprocal, and that the village never takes more from the living land than it returns to it—not just materially but with prayers, propitiations, and praise.[1]

Today the world has become deeply out of balance. Part of our role as contemporary Witches and healers must be to restore that balance, and to do so, we need a deep understanding of how both physical and subtle energies work.

PHYSICAL ENERGY OBSERVATION

In your home base, breathe and ground and come into your senses. Now observe the energy exchanges around you. How is energy coming into this place? Who is its first user? What are the subsequent tiers on the cycle? Where is the energy being stored? How does it leave this place? How many levels of energy transformation, from sunlight to plant to animal to predator, can you see around you? How much of the available sunlight is being used?

Every transformation of energy and matter uses energy. When a plant transforms carbon dioxide and water into sugar, it uses energy from the sun. When a deer eats that plant, she gains energy from it but expends some in finding, grazing, and digesting the plant. Abundance, for the deer, means expending less energy to get food than she gets from the food she eats.

Because each transformation requires energy, some energy is used up as we climb each tier; less energy is available, and therefore each tier supports fewer individuals. There are far more grass plants than there are deer, for example, and far more deer than cougars and coyotes. The tiers of transformation form a pyramid, with its base the plants that directly convert sunlight, and its top the predators.

Some of the energy expenditure in the deer's grazing goes to maintaining the basic life processes and existence of the animal, beyond the energy needed at the moment to browse that particular plant. Every organism, and every thing created by human beings or other organisms, embodies a certain amount of energy.

As I look around the forest that surrounds this cabin, I see an enormous amount of embodied energy. Redwoods tower more than a hundred feet into the air, filling the space around me with wood and needles. Birds flit through, jays dive-bomb the cat food, squirrels leap from branch to branch. I'm sitting in a cabin built of wood and glass and plasterboard, with a cast-iron woodstove and a propane-powered refrigerator, and with chairs and a bed and books and a computer, all of which represent "embodied energy"—the energy and materials used to produce these things, transport them, and (in the case of machinery) get them up and running.

The energy embodied in the living things around me is called "biomass," the sheer weight and substance of biological life in a system. In this forest, the biomass extends up for a hundred to two hundred feet, and downward into the ground in the form of roots and the organic life of the soil. In an old-growth redwood forest, the biomass might extend three hundred feet above the earth.

Energy is embodied in this system, and also stored, held available for future use. I could, in theory, cut down the redwoods and burn them in the woodstove, releasing as heat and light some of the sun's energy that grew them. I won't do that—but I will thin the tanoaks for my winter fuel. I can take as much wood out of the forest as I need or want, without hurting it in the least, as long as I remain within the solar budget.

The solar budget is the amount of the sun's free energy transformed and stored by the forest each year, beyond what it needs to maintain itself and what it uses up in making the transformations. It's the only true profit margin that exists.

I can increase real abundance by increasing the amount of sunlight that is put to use. Compare an acre of redwood forest to an acre of parking lot. In the forest, trees three hundred feet tall are madly using that sun to turn air and water into needles, twigs, and wood, feeding billions of organisms. On the asphalt, all the sunlight is doing is creating heat and glare where they're not wanted. If I'm renting out the parking spaces to shoppers, the lot may produce a form of monetary abundance for me, but in an ecological accounting, paving that space is wasting a huge potential resource.

The key to sustainability and real abundance is to remain within the solar budget. As Mr. Micawber wisely says in the novel *David Copperfield*, "Annual income twenty pounds, annual expenditure nineteen nineteen six, result happiness. Annual income twenty pounds, annual expenditure twenty pounds ought and six, result misery."[2]

As soon as I exceed the solar budget, I start to degrade the system. Were I to clearcut the redwoods, for example, I'd be removing many years' worth of embodied energy. As well, the trees represent water, nutrients, and minerals from the soil. In a pristine forest, the redwood outside my window might grow for a thousand or two thousand years, but eventually it would fall and die, returning all those minerals and nutrients to the soil in a slow process of decay that might also take a thousand years.

If I remove all that biomass and stored energy, those nutrients and minerals, from this forest, it might still have enough resources to grow back over time. But in effect, I'd be using up many decades' worth of stored energy in a short time. I could certainly get away with this once, maybe even twice or three times (if I left long enough intervals in between). But over time, the forest would lose fertility, resilience, health. And indeed, that is exactly what has happened to this forest, which was logged at least twice in the past hundred years, long before I got here. The redwoods are still growing, still beautiful, but when I compare these woods to a stand of old growth, where trees the size of the stumps around me are still alive and flourishing, I can see just how much has been lost.

When we use fossil fuels, we are using up energy that was stored millions of years ago, and that cannot be replaced in anything short of geologic time. The U'Wa people who live in the cloud forests of the Colombian Andes have been resisting oil drilling on their traditional lands by a U.S. oil company, the L.A.-based Occidental Petroleum. They believe that oil is the blood of the earth and should not be disturbed.[3] Fossil fuels are an incredible gift from Gaia, a huge reservoir of potential energy, which we are squandering daily, at the cost of enormous pollution and destabilization of the world's climate. A sane energy policy would move us as quickly as possible toward reliance on renewable sources of energy. As part of our shamanic responsibility of maintaining and restoring a balance, we should be both advocating for such policies and doing what we can as individuals to reduce our expenditure of the earth's blood.

A Simple Energy System

The energy flow in living systems is huge and complex. To understand it better, it might help to look first at a relatively simple energy system—say, the one I'm using right now.

Here in my cabin, on this hot July day, my computer is powered by solar panels that directly convert the sun's energy into electricity. The solar panels, the batteries, and the whole electrical system represent a certain amount of embodied energy. But once in place, they make direct use of the sun's gift. Not only that, they are pretty reliable and relatively trouble-free, the batteries being their weak point. Batteries are like pets—they need to be kept fed (or charged with energy) and watered, and they can't be allowed to get too hot or too cold.

My solar electrical system requires some of my own energy and attention as well. It takes more observation and consciousness than the grid electricity most of us are used to. During the day when the sun is out, the panels are producing electricity and storing it in batteries. I can sit here and write all day on the sun's bonus and still fully charge the batteries that will power my lights and computer and maybe even a video or a CD at night.

But at night, I have only the energy stored in those batteries available. In fact, if I want them to last, I can't use more than the top 20 percent of the energy they store. So I must watch what I use carefully. In the city, I can leave lights on and the stereo running and the TV going in another room and it costs me just a few dollars extra at the end of the month; *here* it might cost me a $500 set of batteries. So I carefully monitor what I use and how full those batteries remain. In the fall, especially, when down here among the redwoods we don't get a full day of sun, my supply is marginal. When the batteries get too low, I must stop using power or find another source.

In the winter, when the rains come and our stream is running, we have a micro-hydro system that provides our electricity. We catch water upstream in a tiny dam not much bigger than a milk crate and run it through a "race," or pipeline, to a spot below the cabin, where it turns a wheel that generates electricity. Then the water goes back into the stream. We "borrow" the water; we do not remove it from the ecosystem.

The micro-hydro, when it's running well, puts out five or more amps of power—not a huge amount, but enough for our needs. Without going into the technicalities of electricity, on our twenty-four-volt system that's enough to run a laptop, power a few compact fluorescent lights, and play a CD or watch a video. Because the micro-hydro runs twenty-four hours a day (and has the batteries as backup), it provides more than enough electricity for our needs—so much, in fact, that the system includes a shunt to divert excess power lest it overcharge the batteries and burn them out.

You could say this system has three basic parts. First, there's the *input*—the energy that comes from the sun or the flowing water. How much of that we take in depends on how many panels we have, or on how much water is flowing in the stream, what volume of water we collect, and how much pressure the water is under when it turns the wheel. We could increase abundance, or available energy, by increasing the number of panels or the volume of water we collect.

The second part of our system is *storage*, which provides a buffer. The batteries allow us to collect excess energy when we have a lot and to make more than our input available when we need a lot. We could also increase available energy by increasing our storage capacity. In summer, when the sun is high, and winter, when rains fill the stream and our hydro system is working, we undoubtedly produce more energy than we currently store. A second set of batteries would effectively make twice as much energy available to us. But in fall, when the hydro stops running and the sun is too low to fully charge the batteries we have, more storage wouldn't help us. And because batteries need to be kept charged to maintain their lives, having more than we can keep fully fed all year might become a liability.

The third part of our system is *output*—the use the energy is put to. We can also create more abundance by reducing our output—reducing, for example, any use that doesn't directly contribute to our abundance and quality of life. Turning the lights off when we're not using them, eliminating the "phantom load" of appliances that draw small amounts of power even when off, using compact fluorescents instead of incandescent bulbs—all these steps make more energy available for what we truly need and want.

The intake, storage, and output of energy in my system are regulated by several little gadgets that include some form of self-regulating cycle. For example, when the sun has fully charged the batteries, the charger shuts off so that they don't overheat. When the batteries are drawn down, the device again allows the sun to charge them.

Not everybody is going to run out and install an alternative electrical system, although doing so might be the best investment you could make if you own your home. In California, the state has offered a rebate on systems that are connected to the grid; the rebate pays nearly half the cost of the panels. Combined with the current low interest rates in 2003, when we began this project, our collective in the city was able to refinance the house, lower our payments, install solar panels, and still have money left over to improve insulation and help reduce our energy use.

Generally, conservation (or reducing waste output) is the easiest, most efficient, and least expensive way to generate more abundance. Increasing input would require adding more solar panels or "borrowing" more water. Increasing storage would mean adding more batteries. Both involve additional embodied energy, costly to the planet and to the bank account. But decreasing waste often requires no new material or infrastructure, no more embodied energy.

We don't all own homes or have the option of installing solar panels or a hydro system. But we can all conserve energy. Often we don't because we don't think a single lightbulb makes a difference, or because we remember our parents nagging us to turn off the lights and we still feel resentful decades later, or because we are unaware and unconscious.

But as Witches, we need to be aware of the energies around us, on every level. Paying attention to the amount of electricity we consume will help us develop awareness of the more subtle energies around us. Since part of our role is to be guardians of the balance, we will have more personal power if we cherish the balance even in small ways in our own lives. For personal power derives from integrity—the unity of our actions with our values, our speech with our acts. Each lightbulb left on represents, in a sense, a small leakage of our own personal power. Every time we turn off one of those unused lights, we are building our own store of power.

Subtle Energy

From my journal:

I'm at my home base spot, trying to observe the energies, but I'm being distracted by Tigers and Bears, the left-behind cats of a neighbor who recently moved out of the area. Originally there were three of them, and because one was gold, one was tiger-striped, and one was just big and cuddly, I called them Lions and Tigers and Bears. I've been feeding them, but they want more. Tigers is butting my legs and purring until I take him up on my lap to stroke and cuddle and scratch behind the ears. I try to brush the burrs out of his fur, but he butts my hand and the brush, just wanting to be petted. The cats aren't hungry—there is food left in the bowl. Tigers just wants attention, energy, love.

Besides the physical and electrical energies that science can measure, there are other, more subtle forms of energy. Witches know that this energy is as vital to life as the more tangible forms. We all need love, attention, and energetic support to thrive.

Many other cultures have words for this energy: the Chinese call it *ch'i*, the Hindus *prana*, the Hawaiians *mana*. Many non-Western healing traditions are based on the understanding of subtle energy and how it impacts our physical and emotional health.

Witches have also worked with these energies, for healing and in magic. We can learn to observe and shift subtle energy—indeed, that is a basic core practice of magic, a part of opening to the wider perceptual world around us.

SUBTLE ENERGY OBSERVATION

Conduct this observation with a partner. You can also do it in groups, by pairing off.

Sit with your partner for a moment. Hold hands and match your breathing. Slowly let yourself become aware of how much you know about this person, just by being with her or him. What information are you receiving? How? Which senses are you using? What subtle senses or feeling beyond the physical come into play?

Now open your eyes. Gaze into your partner's eyes and become aware of all the information you are now taking in.

Close your eyes, and return your focus to yourself. What changes?

Now bring one hand up in front of your mouth and feel your breath. Bring your hand out to the edge of your breath. Become aware of how subtle the edge is where your breath becomes part of the overall air. It may be just a sense of heat or moisture.

Opening your eyes, turn your hand around to feel your partner's breath. Again, move your hand out to find that subtle edge.

When your hand is at the edge of your partner's breath, you'll be at about the edge of her/his energy field—what we call the "aura." Continue moving your hand around until you can feel that subtle edge around different parts of your partner's body. Again, what you feel will be subtle, a sense of heat or tingling or just an urge to stop at a certain place.

Take time to explore each other's auras, to notice where the energy is stronger or weaker, hotter or colder, to be aware of any emotions or images that arise for you. Share what you perceive with your partner, to check if your perceptions make sense to her/him.

When you are done, shake out your hands to release any energies that may still be clinging to you. Close your eyes, say your own name, and think of five physical differences between you and your partner.

ENERGY BRUSHDOWN

A brushdown is a very simple way to do a quick energetic cleansing, perhaps to release negative energies you have picked up from others around you.

If you are working with a partner, first chop up your partner's aura by moving your hands swiftly through it. Then use your hands to pass through it repeatedly, combing out and flicking out any energies that don't belong. Then fluff, using a motion somewhat like fluffing up a curly hairdo. Your partner can then repeat the process on you.

You can also do a brushdown on yourself, if possible over running water to carry away any negative energies. I often do it in the shower or, in a pinch, over a toilet. Shake out your own hands after.

Volumes have been written about subtle energies. When you explored your partner's aura in the subtle energy observation, you may have noticed areas of greater energy density or concentration. These energy vortexes, or chakras, are key in many systems of healing and meditation. Many good resources exist for those who want to explore them further.

But the most important magical teaching about subtle energies is that energy follows intention. To direct energy, we must know where we want it to go, what our goals and intentions are. To accomplish anything, whether it is creating a ritual or changing the world, it helps to begin with an intention, and to monitor whether the energy you are putting out actually follows that intention.

Remember your sacred intention from Chapter Two? Just by articulating that intention, you have already begun to bring your energy into alignment with it. Now let's look at that intention from an energetic point of view.

SACRED INTENTION AND ENERGY EXERCISE

If you do this exercise with a partner, one can monitor the other partner's aura as she or he goes through the exercise. Then switch roles.

In a quiet place, ground and come into your senses. Close your eyes and think about your sacred intention. As you do, notice the quality of your own energy.

Remember your sacred intention and state it to yourself.

Think about a choice or decision you made recently that was in alignment with your sacred intention. How did you feel inside? As you remember that feeling, what happens to the quality of your energy? Is there a symbol or image or word or phrase that you can identify with that quality, to anchor and recognize it?

What was the result of that choice or decision?

Now think about a choice or decision you made recently that was *not* in alignment with your sacred intention. How did that feel? How did you feel inside? As you remember that feeling, what happens to the quality of your energy? How is that feeling different from the feeling that the previous choice inspired? Is there a symbol or image or word or phrase that you can identify with that quality, to anchor and recognize it?

What was the result of that choice or decision?

What options did you believe you had at the time?

Now use the magical tool of anchoring to your core self/baseline (from Chapter Four)—that combination of image, touch, and word or phrase that brings you to your neutral, baseline state. When you are grounded and anchored, think about the options that you felt were available to you. Do they look any different now?

Now think about the ways you are using energy in your life right now. Which are aligned with your sacred intention? Which are not aligned?

Is there a decision you are facing right now? Think about your options, each choice you might make. Does it feel like alignment or nonalignment? Does it evoke either of your earlier anchoring images? What happens to your energy if you make this particular choice?

Breathe deep, open your eyes, and relax. If you've been working with a partner, switch roles and go through the exercise again. Monitors should be sure to shake out their hands after monitoring, to release any energies they may have taken on. Now take some time to discuss what you felt and observed.

Use this exercise to help you monitor your use of energy and your evaluation of choices. As you bring your energies and decisions into alignment with your sacred intention, notice what shifts in your life, how much energy you have available to you, and what drops away. Keeping a journal can be very rewarding in this process.

Form Follows Energy

Another important magical teaching is that form and matter follow the flow of energy. In our materialistic culture, we're given the message over and over that our energetic and emotional state is conditioned by what things we possess. "If only I had a nicer kitchen," we think, "I'd stay home and cook dinner and we'd eat together as a family and feel closer."

Magic teaches us to use our "will"—our ability to consciously choose to act and direct our energies toward our intention. If our intention is to feel closer as a family, then we would choose to stay home and cook even in our old and grungy kitchen. Being there more, we might clean it up, repaint it. With the money we saved by cooking at home, we might be able to afford some new appliances or a window in that dark wall. The form of our kitchen would begin to reflect our intention and the flow of our energy.

Will is an empowering concept. It teaches us that we can make change in our material circumstances, if we put our will, intention, and energy into alignment. Sometimes that requires faith: we can't always foresee how material circumstances can change in the direction we want—and maybe they won't. But *we* will have changed, and be acting in closer alignment with what we truly value.

That kitchen might remain dingy, but we have a closer, more harmonious family.

Will is not willfulness, not whim, not mere intention to get our way. It's our ability to act "as if," to set out even if we can't foresee every step of a journey. The more we exercise our will, the stronger it gets; and the stronger our magical will, the more we are able to serve what is sacred to us, thereby realizing our deepest dreams and desires.

AWAKENING WILL EXERCISE

Consider again what is sacred to you, and how you are using your energy. Now think for a moment about your current wish list—those material conditions or objects that you think will improve your life.

Is there something you want or desire that you have unconsciously made a condition for aligning your energies and actions with your sacred intention?

Is there some act you could take, some shift you could make, toward your intention regardless of your current circumstances?

Breathe deep, focusing on your solar plexus, the energetic seat of will. Imagine filling that center with a beautiful golden light.

Say, "I honor my magical will, my ability to choose. My will is strong, and I choose to do ________________ to further my sacred intention."

Now *do* it. Do it regularly, even for short times and in a small way. Instead of saying, "Oh, if only I had money and didn't have to work, I'd write," write every day, even if it's only a page in your journal. Do it for at least a full moon cycle. Then see if any of the circumstances of your life have changed.

ASSESSING SUBTLE ENERGIES IN NATURE

Plants, trees, animals, and whole ecosystems have auras, or fields of subtle energy, just as people do. When you've become familiar with a human field, turn your attention to the energy around you. Take some time and feel the energy fields of trees, plants in your garden, your cat or dog. What do you notice about them?

Each time you do the meditation of coming into your senses, include your subtle senses, that ability to sense subtle energies.

How do the subtle energies differ in the city? The wilderness? The forest? By the ocean? What happens to your own energy field in these places?

Like individuals and single objects, places have energy fields as well. In your home base, close your eyes and extend your subtle awareness out, not to any specific object but to the whole of the place you are in. What does it feel like? Just as each person has a distinct odor that a dog can smell, each place has its own energy signature. Is there a symbol, a pattern, an image, a word that describes the quality of this place? How would you recognize its energy?

Open your eyes, come into your senses, and with eyes open, extend your more subtle awareness. Does anything change?

HOME ENERGY INVENTORY

Electricity and other forms of overt energy have an effect on our more subtle energies. As we develop awareness of the energies in nature, we also need to become aware of how the energy fields around us may impact us.

Begin in one room in your house, perhaps your bedroom. Look around and notice everything in the room that uses energy: an overhead light, a lamp, a heater, a clock radio, a computer. Where does that energy come from? What does it cost? In money? In environmental impact? Does the energy's production or delivery create waste products or pollution? What use are you making of the energy?

Are there appliances that stay on all night? Do they create sound or light? How much background noise do they produce? Does their sound or light impact your sleep?

Now use your ability to sense subtle energies. Explore the energy field around your computer, your clock radio, a lightbulb when it is on and off. What differences do you notice?

Turn the lights and other equipment on. Close your eyes, stand in the center of the room, and notice the quality of the energetic field these things create. Now turn them off and notice the difference.

Move on to the kitchen and repeat the questions. Have you learned to associate the hum of the refrigerator with food and home? How many lights are still on when you turn everything off? Do you have a microwave always ready for action? Move on to the living room, the home office. Notice the VCR, the stereo, the cordless phone, the fax, all the helpful electronic gadgets that never turn off. Everything that has one of those little black transformer boxes at the plug produces a phantom load.

How do the sounds, the lights, the constant energy use affect the whole field of the room? How do you feel in it?

Now turn everything off. If you can, even unplug the refrigerator for a moment. What changes? What sounds and feels different? How does this affect the subtle energy field of the room? Your own energetic field?

Consider ways to turn more things truly off when they are off. Stereos and computers are best plugged into surge protectors, and the switch can simply be turned off when you're done. If you are rewiring a home or building a new one, the electrical plugs can be wired to a master ON/OFF switch by the door.

As you become more aware and more sensitive, you may find that the continuous electrical stimulation that permeates modern life becomes first noticeable, then irritating. If you spend long hours at work in front of a computer, when you come home you may want at least one room to be a haven where you can turn everything off for a time. Your bedroom is especially important. Protect your sleep. Don't keep a clock radio right next to your bed—in fact, if you can, do away with it altogether and wake up by some other means. If you keep a telephone in your bedroom, make it a simple one, with no black box attached. Turn your computer and printer off before you go to sleep, and charge your cell phone in some other room. Be especially conscious of this if you have trouble sleeping or have an autoimmune disease or other disorder that drains your vital energy. If you think this whole idea is ridiculous, try it for a week or a month and notice what changes.

Becoming aware of energy use is the first step in reducing it. The energy we use at home rarely comes from a benign source. If you don't know how your utility company produces energy, find out. Does it come from burning coal? Damming a river? Running a nuclear power plant? If you know what you are implicated in by using energy, you may be motivated to use less.

Conservation is important, for bringing our own lives more into alignment with our values, for how it can impact the energetic fields of our own homes, and because small changes do add up and make a difference. But for real change to occur, we need changes in public policy. Our collective wouldn't have put solar panels on our house in the city were it not for a public policy that made it financially possible. The more we, as individuals, understand how energy works, and the more we know about alternatives, the stronger advocates we can be for those changes in policy that will help bring us into balance. It's worth burning that light an extra hour, or firing up that computer again, to write a letter to your representative urging more funding for solar panels and less for oil wars. And I will personally absolve you of a dozen left-on lightbulbs if you organize a group to pressure your utility to subsidize solar or other clean power.

As Witches, we are always working with both overt and subtle energies, increasing our ability to be aware of both of them, and eventually learning to shift and change them. We consciously "run energy"—move subtle energies through our own channels and energy field.

That kind of work *uses* energy too, and we need strong sources of the vitality found in nature in order to remain vitalized and healthy. Over the years, I've become concerned as I've seen many Witches, healers, therapists, and other spiritual teachers develop immune system disorders, chronic fatigue syndrome, and other illnesses. In order to work with energy as we do in the Craft or in other healing traditions that derive from indigenous and shamanic practices, we need a strong, ongoing connection to the natural world. These techniques come from people who were immersed in the natural world and attuned to other consciousnesses and communications. Early practitioners of the Craft didn't include in their instructions, "Go outside regularly, garden, and eat local food in season," because they didn't need to, any more than I would think of advising you to bathe regularly. I assume you do—since just about everyone in our culture does. Likewise, in the past just about everyone spent time outdoors in an environment that was more highly vital and less degraded than ours, grew some of their own food, and ate other food that was locally grown and in season because there weren't many other options.

But today we are often working with techniques of moving energy or awakening consciousness that have been removed from their original matrix of a vital, natural world and distanced from their original function of maintaining the balance of energy flow between the human and nonhuman worlds. As David Abram writes,

> Shamanism has . . . come to connote an alternative form of therapy; the emphasis is on personal insight and curing. These are noble aims, to be sure, yet they are secondary to, and derivative from, the primary role of the indigenous shaman, a role that cannot be fulfilled without long and sustained exposure to wild nature, to its patterns and vicissitudes. Mimicking the indigenous shaman's curative methods without his intimate knowledge of the wider natural community cannot, if I am correct, do anything more than trade certain symptoms for others, or shift the locus of *dis-ease* from place to place within the human community. For the source of stress lies in the relationship *between* the human community and the natural landscape.[4]

Abram may be a bit categorical: I actually think many of these techniques can be very useful in human insight and healing. But when they are practiced outside of a relationship with the natural world, they pose a danger of devitalizing the healers as well.

Your best protection, and an important aspect of your own healing, is to spend time in the natural world, preferably in a place that is high in both physical and subtle vitality. Those old Victorians had the right idea when they sent invalids to the seaside for a cure. The more time you can spend in nature, grounded and fully in your senses, opening up to its communications and gifts, the more vitality you will have to draw on. In my own life, consciously adopting a practice of observation and connection with nature has made an immense difference in my health and energy. I have far more energy now, at fifty-two, than I did ten or fifteen years ago, I enjoy better health and fewer bouts of the flu, and I can tolerate physical discomfort, danger, and intensity much more easily.

Besides observing the energy around me, I also consciously use it as a source and take it in. Below is a way to begin that practice.

DRAWING ON A HOME-BASE ENERGY SOURCE

In your home base spot, ground and come into your senses. Now activate your more subtle awareness. What energies can you feel around you? Are there energies that can feed you?

Close your eyes for a moment and feel your grounding cord. Ask the energies and spirits of your place for help and support, and breathe energy up through your roots, filling your body and aura.

Continue to breathe in, imagining that you can absorb some of the energies of the place around you through your aura. If you sense any weak spots or deficient areas in your own energy field, consciously fill them.

Breathe down energy from the sun, moon, or stars—whichever are out. Imagine that energy filling your aura, taking it in through the leaves of your own tree.

Thank your place and its energies, and open your eyes. Do this regularly, at least once a day.

You can also use this same practice away from your home base, in any place of high vitality or whenever you need to renew and replenish your energy.

GIVING BACK

"We take from the earth and say please. We give back to the earth and say thank you," says Julia Parker, a Kashia Pomo.[5]

Since our goal as Witches is to maintain and restore balance, we can't just take energy without returning something to the place and powers that feed us. There are many ways that we can give back.

Gratitude and Thanks. Just expressing our praise, thanks, and appreciation is a way of returning energy.

Prayer and Song. A chant, a liturgy, a song, a spoken prayer are all ways of giving back. Try making up a special song for your place and singing it each time you return, or choose a particular chant you know that seems to fit.

Physical Offerings. Many cultures give back to nature by offering something tangible: tobacco, ti leaves, milk. Try leaving a physical offering of some sort in your home base each time you go there. Be aware that the distinction between an "offering" and "garbage" is sometimes subtle. I prefer sacred water as an offering, as it leaves nothing that can be construed as a mess.

Actions. Actions and activities can be consciously dedicated as offerings. Maybe turning off that lightbulb can be an offering you make to the spirits of the land. You can dedicate an action by saying something like the following: "Spirits of this land, I am turning off this unused light out of gratitude for the gift of life—the sunlight that feeds all things. May the balance be restored. Blessed be."

Energy in Living Systems

Living organisms also store energy to be released as needed. If humans had no means of energy storage, we'd have to eat nonstop just to keep breathing.

We have self-regulating cycles that help monitor our intake and storage of energy. When our supply of food is low, our blood sugar drops and we feel hungry. When we eat, our blood sugar rises and we feel full.

So, too, the larger ecosystems of the earth self-regulate. Those tiers of transformation of energy, from sun to plant to grazer to carnivore, function as one self-regulating whole. On the pristine prairies of the pre-pioneer Midwest, an incredible diversity of plants used the sun's energy to make food. Their biomass extended deep underground, in root systems sometimes twenty feet deep. Herds of bison passed through periodically, grazing the grasslands, removing growth that would otherwise accumulate and choke new shoots, dropping their fertilizing dung, churning up the ground with their hooves so that rain could soak in and new plants could get a start on life. Wolves hunted the bison, culling the herds and taking out weak and sick animals. Fear of the wolves kept the bison bunched together and on the move. The surging throng of huge animals gave the plants and ground the periodic disturbance needed for health, moving on often enough that the bison did not overgraze and destroy the grasses.

Remove any part of this cycle, and you disturb the balance and flow of the whole. The prairie needs the buffalo, and the buffalo needs the wolf. Predator and prey are not at war with each other; they are part of the same system of relationships that are mutually beneficial, if not to individuals, then to the species and ecosystem as a whole.

Many environmentally conscious people choose to be vegetarians or vegans, to eat from the lower tiers of the energy pyramid. This is certainly a conscious and moral choice. The way most meat is raised today, in factory feedlots or horrific pig and chicken concentration camps, is a travesty on both ecological and compassionate grounds, and I try not to eat such meat.

But I do eat meat. I realize that this admission is likely to bring me more irate mail than even my stance on Palestine or the Iraq war, so let me state my case.

I eat meat because in thirty-five years of offering the Goddess every possible opportunity to tell me what to do, she has never suggested I become a vegetarian. On the occasions when for whatever reason I've eaten a vegetarian diet for any length of time, I've become devitalized and sick. Many people thrive on a vegan diet, but many others don't. Our bodies are different. Women, especially those of us who are prone to anemia and low blood sugar, may need a higher protein diet that includes meat. A vegetarian diet has been associated with chronic fatigue, especially in women. Meat is higher in what Chinese medicine calls *ch'i,* or vital life force, and those of us who are working with subtle energies in the way Witches do seem to crave it. Certainly I've been at many vegetarian Witch camps (week-long magical intensives) where those of us who were teaching went to great lengths to make sure we had our own supply of cold chicken in the back refrigerator, just to keep from keeling over with exhaustion by the end of the week.

But let me also make the environmental case. One of the arguments for a vegetarian diet is that it takes many pounds of grain or grass to produce a pound of beef or lamb or chicken. In terms of factory farming, that's certainly true, but I'm *not* defending our current industrial agriculture system.

With two billion hungry people in the world (the argument goes), shouldn't we be using the grain that farm animals eat to feed the masses rather than produce steaks for the rich?

That argument sets up a false choice. First, most of those hungry people are hungry not because food for them doesn't exist, but because the inequalities of our current economic system have left them unable to pay for it, or because the infrastructure and transportation don't exist to get that food to them.

Second, it assumes that land that is used for raising meat could be used, instead, to grow grain or soybeans. But not all land is suitable for growing crops. Drylands, wildlands, steep hillsides, and other areas that are limited in water and access, and that are easily eroded, could be devastated by attempting grain production on them. Indeed, that has happened in much of the original shortgrass prairie lands of the American West. Where I live, a soybean or wheat farm would be an environmental and financial disaster. But my neighbors Jim and Dave raise sheep and goats on a small scale with great sensitivity to the land, and I'm happy to help support their sustainability by buying and eating some of what they produce.

But, you might ask, don't sheep and cattle degrade and erode the land? They do if they are not managed well. However, with sensitive management, they can also improve the land. South Dakota rancher Jeff Mortenson, a student at one of our Earth Activist Trainings, has restored much of his ranch by mimicking with cattle the animal impact of the ancient bison herds. He grazes more cattle per acre than his neighbors, but keeps them concentrated for short times in areas and gives the plants time to recover. Native shrubs, grasses, and streamside vegetation have returned to his land, and its overall health is vastly better than that of neighboring ungrazed government lands.

Jeff follows a system detailed by Allan Savory in *Holistic Management*.[6] I strongly recommend Savory's book, not just for those who are actually interested in raising sheep or cattle, but also for anyone interested in improving overall decision-making. Savory makes a convincing case that land in climates that are "brittle"—that have dry periods when water is less available and organic materials do not break down easily—require animal impact for their health. Overgrazing will harm them, certainly; but so will lack of any grazing. But pastureland can be restored and revitalized by animal impact that mimics the effect of large herds moving in response to predators—the pattern with which grasslands coevolved.

Much of the hunger that exists today in the world is located in just such environments—brittle climates where the process of desertification is progressing rapidly. Contrary to the vegetarian argument, ending hunger might actually require meat production, but only production carried out in conjunction with nature and with respect.

Of course, many people choose to not eat meat because they don't want to kill another sentient being. I am certainly not going to argue with that choice, but my morality is different. Sheep, cows, chickens, and our other domesticated animals have now coevolved with human beings for ten thousand years. Were we to stop eating them, they would not be living happy, productive, fulfilled animal lives; they would cease to exist as species and become extinct. They could no longer survive in the wild. Most have lost the ability to protect themselves against predators and to find food without human help. Others, such as pigs that revert to a feral state, have a devastating impact on the environment. Yet our domesticated animals are also a precious source of genetic diversity, in the thousands of local breeds that have been developed for specific conditions over thousands of years. Their mass extinction would be a tragedy. Humans and our domesticated animals now form a whole, in which we function as top predators, and sheep and cows and chickens need us as we need them.

That doesn't give us the right to confine them in cages for their entire lives, or to condemn them to a miserable existence in meat factories. But if we raise them in humane conditions—conditions that give them a chance to do the things sheep and cows and chickens like to do, that integrate them into a healthy environment and help them perpetuate their kind—if we make sure that they are killed with a maximum of respect and a minimum of suffering, and if we give gratitude to their spirits for their lives, then I am happy to eat them. Death is a part of life, for all of us. It cannot be avoided, for humans or chickens or pigs. When my time comes, I would rather be eaten by a cougar or disposed of by vultures than embalmed and stuffed in a box. But however I am disposed of, something will eventually eat me. And so the cycle goes on.

Loving the Body

What we eat is tied to how we treat our bodies. The Goddess traditions celebrate the body and value life in this world. Our fleshy, living, breathing, eating, and excreting bodies are sacred, the matrix of form for the unique energies we each are. Treating our bodies well is one way we honor the Goddess and show respect and gratitude for the gift of life.

For many people, self-care around food seems to focus on what *not* to eat—meat, wheat, dairy, etc. But to increase the energy we receive from our food and strengthen our own vital energy, we need to think about what we *do* and *should* eat.

Some years ago, when a group of us had just acquired land in the coastal hills of northern California, I was meditating in our garden. I was questioning my own driving need to garden in an environment that was already beautiful and needed no improvement. The garden said to me, "Grow food. I want you to grow food—because if you eat the food that comes from the land, you will *be* the land."

When we eat something, we literally take in the minerals and energy of the place where it was grown. In an indigenous culture, almost everything people ate came from the land they lived on. Their bodies were literally made of the same stuff as their land. People downriver or over the hill would have smelled different. Myth and religion reflected this close identity. In Mayan mythology, for example, people were made of corn.

Jeanette Armstrong, Okanagan writer and teacher, writes,

> The Okanagan word for "our place on the land" and "our language" is the same. This means that the land has taught us our language. The way we survived is to speak the language that the land offered us as its teachings. . . . We also refer to the land and our bodies with the same root syllable. This means that the flesh which is our body is pieces of the land come to us through the things which the land is.[7]

To eat, then, was not just to take in a set of chemical nutrients. It was to be in profound relationship with a place—with the energies, elements, climate, and life community of that spot on the earth—to ingest the place and become it.

Everything you ate would also be something you had a relationship with, that you had yourself grown and tended, gathered, or hunted. All those activities would be sacred—that is, highly valued and marked by ritual and prayer and ceremony, by offering gratitude and respect for the life-forms you were culling. All food would be harvested, hunted, and prepared by people who were consciously putting themselves into a thankful and loving energetic state, and the food itself would carry that energy.

Today, few of us have that kind of relationship with place. The food we eat has often been transported halfway across the world. It may have been grown and picked and processed by people who were exploited and suffering. It may have been doused with poisons, grown in dead and devitalized soil, irradiated. If it is factory-farmed meat, we are eating confinement and torture with every mouthful. If it is fast food from a chain, it may have been prepared by underpaid, exploited, and resentful help under factory-like conditions. Overall, far more energy will have been consumed in its production and transportation than we will get by eating it.

As restorers of the balance, and as lovers of our own bodies, we need to be conscious of what we eat. The foods that we need to maintain our own vitality are also what will help restore the balance. Rather than thinking about what we shouldn't eat, let's think about what we can and should eat to increase our health and vitality.

First, we can eat high-vitality food. This means food that is organically grown, minimally processed, and locally produced (so that it doesn't lose its vitality on its way to you). If you think you can't afford to switch to an all-organic diet, think about what you can add rather than feeling guilty about what you don't want to give up. Can you substitute one organic fruit or vegetable a week? Can you take a bit of extra time to go to the farmer's market, where organic food is cheaper and you are also supporting local small producers? Can you join a local CSA, which stands for Community Supported Agriculture[8]—where for a low price you directly support a local farmer who will deliver a box of fresh fruits and vegetables each week?

We can primarily eat food that is in season. Such food is generally cheaper, can be grown locally, and keeps us in touch with the cycles and changes of the year. We can eat peaches from our garden or the farmer's market in summer, not Chilean peaches in the dead of winter. We can eat asparagus in the spring, and apples in the fall. Of course, if we live in an area with cold, snowbound winters, we will have to import some of our food, but we can still choose to eat locally through many seasons of the year.

We can grow some food of our own. Even if we grow only a small amount, growing and eating food from our place will establish a relationship with it. If nothing else, grow a few herbs or some mint on a windowsill, and use them in rituals and for tea. If you have never gardened or think you have a black thumb, consider that there are basically two kinds of plants in the world: the ones you can't grow and the ones you can't kill. Find some of the ones you can't kill in your area, and grow them. Many of the herbs and medicinal plants fall into this category, as they derive from weeds and by nature have high vitality and resilience. Ask some advice from your gardening friends. And if you can grow some of your own food, you'll find that the vital energy in a salad picked fresh from your garden is far greater than anything you can buy. The taste is also incomparable.

Cook and prepare food with love and gratitude, as a conscious meditation and offering. Karin, who heads the cook team in our Earth Activist Trainings and at the California Witch Camp, is a Buddhist who puts love, compassion, and joy into all the food she prepares. The kitchen is always a happy place to be, and the cooks are generally dancing to good music as they chop and stir. When we eat Karin's food, we feel nourished spiritually and emotionally as well as physically.

COOKING AND EATING WITH GRATITUDE

Cooking can be an action you offer up: "I give gratitude for the gift of life, for the lives of all the creatures in this food. I offer up this work of preparation as a gift of love and nurturing to all who will eat it, and a gift of thanks to all that has provided it. May the balance be restored."

Stop and give thanks and bless the food before you eat it: "I give gratitude for the gift of life, for the sunlight and the plants that use its energy to turn air and water into food, for all the beings who gave their lives to this food, to all who grew and tended and picked and transported and prepared and brought it to us. May the balance be restored. Blessed be."

Ground and come into your senses before you eat. Eat consciously, savoring the taste and noticing the energy of what you take in.

One of our primary energy storage systems in the body is fat. As humans, we have laden our flesh and fat with huge symbolic meaning. In many early Goddess cultures, as in Hawaii and Polynesia today, fat is seen as desirable, a symbol of power, a literal sign of stored energy. The earliest representations of the Goddess, from the old Stone Age, show fat women with huge breasts, bellies, and buttocks.

But for women today, in our fat-phobic culture, loving our bodies can be difficult. All around us are messages telling us that fat is ugly, undesirable, unhealthy, and a sign of weakness of character. The standard of beauty is lean and mean, and even our natural curves and softness are suspect.

Fat can certainly be a sign of imbalance—of a diet that is not truly nourishing, of lack of exercise and physical vitality. If you don't exercise, you should find some form of physical workout that fits your body and your routine, not to lose weight but to maintain the high level of physical vitality you need to work with more subtle energies. If you are not eating vital, nourishing food, you are putting your overall health at risk.

But fat can also simply be how your body metabolizes and stores energy. If you are vital, active, healthy, and careful about what you eat, and are still fat, that may be the way your body wants to be. Like redwoods, as we get older we get thicker. Hating our bodies, depriving ourselves, judging ourselves, trying out one diet after another—these are far worse energy drains than leaving on the lights. Instead of those negative perspectives, try the following meditation.

BODY LOVE MEDITATION

(Especially for Fat Women.)

(Okay, Fat Men Too.)

(Oh, all right, you don't have to be fat to do this.)

For this exercise, you need a full-length mirror and a bowl of salt water. Use only a few grains of salt, as you will eventually give the water back to the earth.

In a private place, ground and come into your senses. Take off your clothes, and look at your body in a mirror. Feel its curves, the fullness of your breasts, the roundness of your belly. Become aware of all the feelings that arise in you. Do you love your body? Hate it? Admire it? Are you ashamed of it? Want to change features of it?

Just acknowledge honestly all that you feel and think about your body. Ask yourself, How much of my energy is tied up in feelings or frustrations about my body?

Place the bowl of salt water at your feet. Breathing slowly and deeply, imagine all that energy, all of your thoughts and judgments, draining off of you into the bowl.

Now look at your body again. Going slowly, from your feet up to your head, look at every part that you might have felt judgment about, and bless it. See your curves as stored sunlight. Acknowledge and bless the stored energy in your body. Thank the plants and animals that transformed that energy into food for you. Thank the sun.

When you are full of gratitude, lift the bowl of salt water and breathe into it. Imagine filling it with sunlight and gratitude, transforming the energies you've poured into it.

When you are done, take some of the transformed salt water and anoint any places on your body that you need extra help in blessing. Then pour out what is left onto the earth. (If you are worried about too much salt affecting plants, dilute the solution even more before pouring it.)

Fire Ecology

In addition to the sun's fire and the subtle energies, fire itself is a purifying and renewing force—and also a destructive force. Fire is the first tool human beings used to alter the environment around them on a large scale.

Fire can be an energizing and renewing force in a landscape. Here in northern California, we live in a fire ecology. Native American tribes regularly burned the forests and grasslands, to keep them healthy and diverse.[9] Fire creates a mosaic of habitats. A wildfire burns at different intensities over different patches of ground, so when it passes through, it leaves a variety of different habitats and stages of succession intact. In southern Oregon, at an Earth First! gathering last May, we camped on the edge of a recent burn. Below us, we could see patches of scorched earth, patches of undamaged trees, lightly burned areas, and areas where all was black and dead. Each would provide slightly different conditions for plants and birds and animals, increasing the area's diversity and the extent of its "edge."

When our forests were burned regularly by Native Americans, they remained open and fires stayed low and relatively cool. In fact, regular burning prevented the enormously destructive wildfires we often experience today. Fire kept the meadows open, so the deer and elk could find good grazing. Burning also killed destructive insects and disease organisms, rejuvenated many plants (providing long, straight new shoots for basketweaving), and fertilized the soil.

But fire is destructive as well as benign. Even on pristine lands, fire lays bare the soil, increasing the likelihood of erosion. Fire can transform massive amounts of biomass into dead sticks and ash, leaving an area with a reduced capacity for transforming the sun's energy. And fire, of course, can easily get out of control.

So fire is not a tool we want to use lightly or automatically. In our part of northern California, where houses now dot the landscape and the forest is burdened with a huge fuel load, we can't easily or carelessly reintroduce regular burning. We are even careful not to use our woodstoves on a cold day in the dry season. Wildfires are a serious danger all summer long, because they ignite so readily. Fires have been started near us by a mower blade hitting a rock and setting off a spark, and by an abandoned bottle of water acting as a magnifying lens in dry grass. Fire is truly our teacher.

The Lessons of Fire

Fire teaches us awareness of what we do and what we leave around. Fire teaches us responsibility and mutual dependence. Fire teaches us nonattachment: we all know that we could lose our homes and possessions at any time, and it's difficult to get fire insurance up here.

Fire also teaches us about community. Our fire protection comes from a volunteer fire department, as it does in most rural areas in the United States. Men and women in the community volunteer many hours each week, without pay, to be on call in case of fire or medical emergency. They also devote many unpaid hours to trainings and meetings. If someone up here gets hurt, it's one of our neighbors who will respond. Those of us who aren't in the fire department also volunteer time and energy to raise money to support it.

Though fire can be wild and untamed, the hearth fire is a great symbol of home. It's where we cook our food and share our meals. It's the center of the fire circle, around which the community gathers for ritual, for council, for storytelling.

As we learn to work in community, fire also teaches us about our own passions and emotions, about sustaining anything that requires excitement and enthusiasm. To learn how to start a movement or spark a creative project, practice building a fire.

Building a Fire

A fire needs three things to burn: fuel, heat, and oxygen. And those things must be present in the right relationship to each other. The Creighton Ridge Fire of 1978 began on a day when the ambient heat was so high, the fuel load of dead grass so dry, that a spark from a mower ignited the grass. On a cool day, that would not have happened.

If you've ever hastily tried to build a campfire or light a woodstove and failed to get the fire going, you know something about what is necessary to start any project. Consider:

What is the need you are trying to address, the problem you are trying to solve?

What is your fuel? What is the community, the set of resources, the range of conditions that you start with?

What is your heat? What is the passion, the enthusiasm, the burning desire?

What is your oxygen? What is the fresh air, the idea, the insight, the clarity?

To build a fire, you begin with tinder, something that is easy to light—paper or pine needles or dry twigs. Who in the community will be first to catch fire with enthusiasm? Can you fire up a small group first, who will help spread the idea to others?

A fire that is only paper will soon burn out. To get a lasting fire going, the tinder must ignite the kindling, the smaller sticks chopped into thinner pieces. Once your small group of original enthusiasts is excited, how do you inspire more people to get involved? How do you keep the structure open enough for air—new ideas and insights—to get in?

To burn for a long time, to heat a room or provide a center for a council, a fire also needs some big logs. But throwing a big log on a fire too soon can crush it. At what point is your idea or group ready to expand to a larger base? What must already be inflamed to support a larger scale of work?

As a fire burns, it needs replenishment. As a project goes on, some of the original resources and people will burn out or move on. How are you going to replenish them? What is your source of additional fuel?

Eventually, a fire does burn out. A project runs its course, or a movement enters a different phase. Do you want to bank your fire, to keep a spark alive for another cycle? Or is it important to put it completely out? If so, how do you cool or smother what is left of it?

Good firebuilders, good organizers, know that you cannot rush this process or skip stages. The big logs won't burn unless you've carefully laid the fuel, left room for air, and gotten the tinder and kindling burning well first.

The following meditation (a version of which is described in *The Twelve Wild Swans*[10]) is designed for use with a group.

FIREBUILDING MEDITATION

Ask all the participants to bring something to make a fire, but don't specify exactly what. When the group gathers, ask each person what she or he brought.

Then ask whether the group as a whole now has what's needed to make a fire. (If you don't, that's the lesson!)

Lay the fire with what people brought, and light it—assuming someone brought matches.

Sit around the fire, and ask each person to consider whether what they brought reflects their role in the community, or (if it's a new group) the role they *generally* play. Did one person bring the tinder, to get a new idea going? Did someone else bring kindling, to nurture it along? Who brought a great big log? And who has ever killed a fire, or a project, by dumping a big idea on it too soon? Is someone the overly responsible one, who brought matches and kindling and tinder just in case no one else remembered to bring anything? Did someone bring nothing at all, expecting that others would take care of everything?

Does the group have what it needs to sustain this fire?

Silent Cheering: Feeding the Fire

George Lakey, a wonderful nonviolence trainer and activist, tells a story about a time when he was training a group of union officials to be trainers themselves. One of his trainees, Joe, had just set his group to doing an exercise. "What can you be doing while your group is doing their work?" George asked him.

"I don't know," Joe said.

"You could be silently cheering them on," George suggested.

"What do you mean?" Joe asked. He was highly skeptical that silent cheering would affect his group in any way. So George had him sit down on the couch and called the rest of the group over to silently cheer for him. He said that within a few minutes, Joe turned beet red and began to sweat, the impact was so strong.

When George told me this story, I recognized a valuable way to explain the concept of energetic support to those who don't think in terms of magical language and resist these woo-woo concepts. Even for those of us who do think in magical terms, cheering can be a useful concept. We cheer for a friend in a race, or for our kids' soccer team, to offer energetic support, a taste of that attention and love that the cat craves. It feels good to be cheered for.

Shortly after I heard this story from George, I was walking through my woods thinking about Sudden Oak Death, a fungal disease that is attacking many of our trees. I was remembering what Mabel McKay, a Pomo healer and elder, used to say—that the trees and plants needed human beings to eat them and use them, to talk to them and sing to them and praise them, or they would die.

Our afflicted tanoaks were one of the favorite foods of the Pomo, but now their acorns mostly go ungathered. Lumber companies think of the tanoaks as "weed trees," something to be gotten out of the way so they don't compete with the more valuable conifers. Because these "weeds" grow back more quickly than the redwoods or firs after a fire or a clearcut, our woods are choked with brushy, many-stemmed saplings that rarely grow into the magnificent grandmother trees the Pomo prized.

It occurred to me that maybe we needed to do a ritual for the trees, to sing to them and praise them, to collect their acorns and make food from them. To begin that process, I decided that on my walk I would try cheering for the trees.

At first I thought this would be an exhausting task. My daily walk takes around two hours, much of it through forest. How could I cheer for every tree? I'd be worn out! But I decided to try. So while coming into my senses and trying to walk silently and respectfully, I was mentally saying, "Go, Tanoaks! Yay, Madrones! Aren't you beautiful today! Let's hear it for the Live Oaks! Hurray! Give it up for the Black Oak over there beside the road." I made up little songs and sang them inside my head, and I collected a bagful of acorns, offering some waters of the world in exchange.

Much to my surprise, when I was done with my walk I was not exhausted, but invigorated. Cheering for the trees had put me in a high-energy state that replenished my vitality and made me feel good. I began to have a deeper understanding of why so many people love spectator sports.

We human beings are constantly looking to one another for energetic feedback for our ideas, visions, and enthusiasms. Cheering someone on is like fanning the flames of their creative fire. It can help that fire burn more brightly, more steadily. When we're in a group or relationship where people are cheering for one another, we become more creative, more joyful, more intelligent. If you've ever had to give a speech in front of a group of people, you may have noticed how even one supportive friend in the audience could bolster your confidence (and conversely, how hard it is to be brilliant and scintillating if people are shuffling their feet, yawning, looking bored, or walking out).

We can also douse each other's fire, either deliberately or out of carelessness. Have you ever been in a group where every idea was met with criticisms and dire predictions of failure? Where everyone who stepped forward to propose a direction got attacked?

In one of our Witch camps, we were working with the story of VasaLisa, a Russian fairy tale about a beautiful young girl who has the proverbial wicked stepmother and stepsisters. At one point, these nasty relatives put out every fire in the house so that VasaLisa would be forced to brave the fearsome Witch Baba Yaga to get fire.

In one nighttime ritual early on in that camp, when everyone was joyfully dancing around the fire, two priestesses acting as the stepsisters suddenly threw buckets of water all over the flames. As the fire hissed and steamed, and we all stood around it in shock, they cried out, "VasaLisa, you stupid girl. You've let the fire go out!"

Throughout that week, we went on to reflect on the ways in which we put out each other's fire. When we recognize subtle energies, we become responsible for the kind of energy we are putting forth in our community. The things we do and say about each other create a subtle energetic field that either supports our work and our relationships, or undermines them.

Malicious gossip, backbiting, unsupportive criticism, and mean-spiritedness douse even the stoutest fire. And because a fire takes energy to build and maintain, such negativity is wasteful of the community's resources; it's like using electricity not just to keep the radio on all the time, but to keep it tuned to an irritating and distracting station.

Anger and conflict don't necessarily douse a fire. Conflict is part of all human relationships—but we can have conflicts, arguments, and disagreements that strengthen rather than undermine the underlying relationships. When anger is directly and cleanly expressed, it can be a bright flame of its own, healing and strengthening a relationship.

But when anger festers, when we chew over our grievances like old bones without expressing them directly, when we meet others with sullenness or resentment, we douse not only their fire but our own.

FEEDING/DOUSING HASSLE LINE

In a group, form two lines facing each other. Reach across and shake hands, so that everyone knows who their partner is.

The facilitator should remind participants of one of my favorite quotes from Gandhi, that when you are trying to change the world, first people ignore you, then they ridicule you, then they attack you, then you win.

Everyone in the first line should ground and center, then think of something they're passionate about, something they'd like to persuade their partner to join with them in doing.

The second line should prepare to ignore, ridicule, and/or attack.

The ground rules are these: interact only with your partner, stay with the interaction and don't walk away, and use no physical contact. Give the group a moment or two to mentally and energetically get into their roles, then begin.

Let the interactions play out for three to five minutes; then debrief the group.

How did it feel to be ignored, ridiculed, attacked? How did it feel to be the ridiculer or attacker? Was there anything the passionate partner did that was effective in establishing real communication? Were you able to hold on to your grounding and passion in the face of your partner's indifference or hostility? What came up for you? Have you faced anything like this in real life? What happened to your energy? To the energy in the room?

Now switch roles and repeat the exercise and debriefing. Then have everyone take a step forward, consciously stepping out of their role, and change partners.

Again, the first line should ground and center, and think about their passion.

This time, the second line will silently cheer for them, lending them energetic support—but without speaking.

Give participants a minute to get into their roles; then go. Let the exercise run about three minutes. After you stop, have everyone switch roles, and then run the exercise again.

How was this different? How did it feel to be energetically supported? To cheer someone on? What happened to your passion or idea in that environment? To your energy? How did the energy in the room feel?

Follow this exercise with an energy brushdown, to release any lingering negativity.

The exercise above is a useful one to do with groups that are starting out, because it can lead to a discussion of what kind of energetic atmosphere the group wants to create. Whether you are in a circle, a coven of Witches, a work team, a political group, or a sports team, people will be more creative, intelligent, and skillful in an atmosphere of energetic support. But if the group is full of people who are ignoring, ridiculing, or attacking each other, no idea can catch fire.

Energetic support, cheering each other on, is one of the simple but powerful ways we can honor the great creative forces in the universe in our own groups. Besides the sun, our creativity is the other essentially unlimited resource on the planet. Whenever we support and feed each other's creative fire, we increase the abundance of the planet.

Rage and Anger

"If you aren't outraged, you aren't paying attention" is a slogan I've seen on bumper stickers. It describes our present moment all too well. As we open our eyes and come into awareness of the great beauty and interconnectedness around us, we can't help but become enraged at the incredible shortsightedness and destructiveness of our current system, and at the violence and injustice that maintain it.

Anger is like fire: it is a vital, life-force emotion that arises when we are endangered or attacked and increases the energy we have available. Like fire, it can cleanse, renew, and heal. But like fire, it can also destroy. And maintaining a high state of rage, over a long time, can burn out all of your resources.

Yet ignoring or suppressing rage can also be destructive. Unexpressed and unacknowledged anger can turn inward, becoming self-hatred or depression. It can escape in unproductive ways, turning us against our friends and allies, instead of serving as the energy we need to change a destructive situation.

Anger and rage can also keep us from taking in information. When we are in a "blind rage," we're no longer looking, listening, or thinking about what we want to do. We're not making conscious choices: we're simply reacting. In that state, we can easily be manipulated or knocked off our balance, physically or emotionally. We may do things that later we will deeply regret.

To be earth-healers, we need to be able to sustain huge amounts of rage and anger, without either losing control or burning out. The first step is to learn to ground even when enraged.

GROUNDING ANGER

If you've been practicing the grounding meditation given in Chapter Five, by now you should be familiar with what it feels like to be calm, energized, fully present, and open.

In a safe space, ground. Focus on your roots going into the earth. Think about something that makes you angry. If nothing in your life is making you angry, you might try this exercise in the morning after reading the newspaper. As the emotion begins to build, feel those roots. Remember that energy can flow down your roots as well as up them. Think of them as hollow tubes, and imagine some of the excess energy of that anger draining down through them, going back to feed the fires below the earth. Continue until you feel relaxed enough to complete your grounding by breathing some of that fire back up, as pure energy that you can use for your own creative endeavors, or to create change in the situation that's making you angry.

You may also need to do something physical to release some of this energy. Try roaring—not yelling or screaming in a way that tears out your throat, but bringing up a deep, solid roar from the pit of your stomach. If there is something in your household that needs to be torn up, chopped, or destroyed, ground first and attack that woodpile or trashbin as a conscious act of energy release. My former therapist used to recommend beating pillows, and while I resisted for a long time, thinking it would feel dumb or forced, I eventually tried it, and the physical act of hitting something was a tremendous release. Whacking a stick against the ground, banging the table with a cardboard tube from inside a roll of giftwrap, conducting a mock sword fight, and smashing a piñata are other ways to release anger physically and safely.

If you are in a situation where you cannot physically or loudly release energy, look for smaller and more subtle ways. For example, I have often embroidered my way through a difficult meeting: the needlework project gives me the opportunity to repeatedly stab something. Drumming for ritual is also very satisfying: it gives me something to hit.

Grounding anger is just the beginning of transforming it, but it is a *necessary* beginning.

The next step is to remember to breathe and ground in the midst of the situation that's making you angry.

IMAGINING A CAULDRON OF CREATIVE FIRE

When you are comfortable grounding your anger, move on to this next step. Feel the anger, ground it, and then bring it back up as fire. Feel it warm your roots, your energy center at the base of your spine, activating all your instincts and energies of survival.

Now imagine a cauldron in your belly, a bubbling brew of creative power. Tell yourself, "This is the fire in my cauldron, the heat of my passion." Feel the fire-energy swirl and dance within you, and build in intensity. Now begin to breathe it up through your body. If there is someplace in your body that needs healing, bathe it with some fire. Breathe the fire up to your heart, and feel it warm and bathe your heart. Breathe it up into your shoulders and out your hands, and imagine it energizing your hands and your power to act and do. Breathe it up into your throat, and imagine it strengthening your voice. Breathe it up into your third eye, and let it strengthen your intuition. Breathe it out the top of your head, like a fountain of fire. Stand for a moment, like a blazing firework, and imagine that all that energy is your creative power, available to you now, a resource for you to use in bringing about change.

More suggestions about transforming rage can be found *in The Twelve Wild Swans.*[11] Rage transformed can be a powerful source of energy, a healing and purifying fire that can challenge injustice and help us renew the world.

BLESSING FOR FIRE

We give praise and gratitude to the fire, the sun's fire that fuels all life on earth, the radiant heat and light that is the source of energy for all beings, from the tiniest alga to the towering redwood, from the grass to the buffalo, from the worm to the hawk. We give thanks for the wildfire that cleanses the land, and we acknowledge its awesome power to destroy and to renew. We ask help in living with the power of fire, that we learn once again how to be in balance with fire on the land. We give thanks for the hearth fire, which warms us in the cold and gives our homes and communities a heart-center. We ask that our hearts be open to learn the teachings of fire, that we understand how to feed the creative sparks that arise in each of us, and that we feed the flames of passion and love for the earth. May the flames teach us how to dance, how to transform our rage into radiant action. Blessed be the fire.

NINE

Water

From my journal:

Water brings the land alive. Water gives the land a voice. As I am writing this morning, I hear the occasional tap, tap of the trees still dripping from the night's storms. All night long—indeed, for days—squalls have been coming in, moving through on the wind, dropping their load of rain, and then passing on. The streams have come alive and their song is the constant backdrop, filling the nights now that the frogs are silent.

I spend a lot of the day watching the patterns in streams. Marija Gimbutas identified the V or chevron as a symbol linked to water. In my stream, I can see the water make V's, flowing fast between a stretch of relatively straight bank, the sides deflecting the currents which intersect each other in perfect series of chevrons, their points facing downstream. The water runs faster, there in the center, and I imagine it must be cutting the channel deeper. In the summer, looking at the streambed, I'll be able to imagine the winter flow.

Friction slows the water on the bottom of the streambed; the upper layers move faster, tumble over. Water moves in spirals. A river is an elongated horizontal whirlpool moving downstream. The shape of this flow is changed by the riverbed, by rocks and logs and obstacles and differences in the underlying soil. But, all else being equal, a perfectly straight, smooth river would still eventually meander, because this spiral flow is like a drill, cutting away soil from the banks on one side, depositing it on the other, creating an S curve. One meander creates another: water speeding up around an outside curve is slowed as it rounds the bed, while water from the inside, freed of its silt, speeds up, and so the curves reverse.

In my little stream, I can see the beginnings of meanders. Years ago, the previous owners had small check-dams built all along the stream, and they work well, creating a more varied bottom to the bed. They slow the stream down, make it hesitate for a moment, spilling silt behind them, and then plunging over in a mini-waterfall that digs a deep hole below. Were this a year-round stream, those holes would create deep pools where salmon or steelhead fry could survive the summer. As it is, they keep silt from moving further downstream, to spoil the gravel spawning beds.

Today, on this bright morning in 1998, some neighbors and I are on a mission to explore another aspect of water. We're trying to measure the drop of our stream to determine whether or not we have enough water pressure to put in a micro–hydro system to generate electricity in the winter. Our stream has a lot of flow when the weather is this wet, but only a very gentle drop where it runs near the cabin.

Building a hydro system isn't simply a matter of sticking a wheel in the stream—those do exist, but they need a much bigger and deeper flow. The system we're considering will collect water and send it through a race of two- or three-inch pipe into a small device called a Pelton wheel, where nozzles will direct the flow over a wheel that spins and generates the electricity. But to work, we'll need twenty-five pounds of water pressure.

Water pressure has to do with drop. There are formulas for these things, but generally, for hydro of this sort, fifty feet of drop is needed. It takes two feet of drop to produce a pound of pressure. We can try to measure the drop with sticks and levels, or we can hook up all the hoses in the world—or at least as many as we can borrow—fill them at the high point in the stream, and put a pressure gauge on the end to measure the pressure. With either method, we'll be able to see how far up we'd have to go to generate enough pressure, how much pipe we'd have to lay, and whether or not the whole thing would be worth it for a few months of power each year, when the rains hold steady. On a year like this, when we've just had to replace our backup generator and we've had rain for eighteen out of the last twenty days, it sure seems worth it.

Water pressure, drop, and flow—all of these things remind me of my junior high school physics class with the hapless teacher whose simplest experiments went wrong. I've avoided the subject ever since.

Water will flow from a high point to a lower point. Put it in a closed channel, such as a pipe, and you can run it up and down in between as long as the end pipe is lower than the beginning *pipe. That's how we can fill a water tank on a knoll and then run the water back down into the valley and up to a second tank behind our cabin.*

We opt for the borrowing-hose option to conduct our test, and my neighbor Ken ends up laying the hose right in the streambed. His son Galen *and I run back and forth, feeding him more hose, bushwhacking down to the stream with hoses until we're able to get a steady flow through.*

Then, exhausted and out of length, we decide to call it a day and finish up when we can borrow some more hose.

In a different season, high summer of 2003:

Today we're going up to the spring to find out why we're getting such a low flow of water. Mer *and* Ken *and I go tromping across the hills, hoping we can even* find *the spring. I haven't been up here much in a while, and the trees have grown. The land looks different. But we do find the spring, in a little draw above a big fir tree—a couple of pipes sticking out from the bank: a shallow depression above. I look for the Water God, a cement sculpture made by my friend Donald Engstrom, but it's gone. Maybe that's why we're not getting much water?*

We dig out the spring box, which is filled with gravel to filter the water. It's a redwood box, about a two-foot cube, with holes in the bottom boards that seem clogged with roots. Inside is a metal pipe with holes drilled into it, also clogged with roots. We ream them out with a coat hanger and a piece of conduit. There is still only a small pool of water in the box, however, and we realize the spring probably needs further work. We'll check the flow down below and come back.

I look at the box, the two or three inches of free water in it. It seems very little to support two households, the gardens, the greenhouse, the fruit trees. This morning we were getting less than a third of a gallon a minute. That's less than five hundred gallons a day—plenty for our conservative personal needs, but not much to support a big garden in this dry land.

And yet, what an incredible gift—clean, drinkable water, straight from the ground.

I walk back across the hills, past the old stock-watering pond that remains from when this land was a sheep ranch. The water is low—lower than I remember it being this time of year. It's clear, above plateaus and channels of mud where tadpoles are swimming. I'm surprised to see them so late in July—they must be fruits of a later mating. Dragonflies and damselflies hover above, the sticklike bluets, the bright red/orange flame-skimmers, and a pair of big black and white ones almost as large as hummingbirds. In the water, tiny snails crawl and a variety of bugs swim. I see something that looks just like a fly flying underwater, using its wings as propellers to move along. I've never seen that before!

On top of the water, the water-striders skim along, supported on the surface tension. Sedges grow along the edges of the pond, and green grasses, as well as lots of pennyroyal that smells pungent as I walk on it.

Below is the dry streambed, where in winter water tumbles down, eroding the steep hillside. When people built this pond years ago, they changed the flow of the stream, pushed it down a different side of the hill. Since then, it's been digging its own channel, pushing down silt that has filled up my check-dams further downstream and that clogs my now-finished hydro system in winter.

This stream flows into Camper Creek, which flows by my house, then turns north to flow down to join Carson Creek and MacKenzie Creek and make its way into the South Fork of the Gualala River.

On the other side of our land, the water runs down to streams that flow east and south, to join up with the South Fork of the Gualala before it makes its bend to head north, following the ridge lines thrown up by the San Andreas fault—lines that keep it from heading straight to the ocean.

Water and Awareness

Meditating by a lake one day, I heard the water say to me, "All water is one—one whole, one awareness. All water is continuously aware of all the other water in the world."

That insight profoundly changed my relationship to water. Instead of thinking of it as a physical substance, I began to perceive it as a flow of life-giving awareness, constantly cycling through the world. To be a Witch, to be someone who has consciously accepted the challenge of serving the powers of life and balance, we must bring ourselves into right relationship with that pervasive consciousness. Only through a balanced relationship with water can we have abundance and thriving life.

And water *knows*. Water spirits, water Goddesses and Gods, however we want to name that intelligence that is so different from ours—*something* knows and feels when we approach with love and respect.

Waters of the World

Many years ago, I asked my friend Luisah Teish, a Yoruba priestess, what I could bring back for her from a trip I was about to take. "Just bring me some water," she said. "I collect water."

I started collecting water for her, and as collecting often goes, once I started collecting it it began to seem valuable to me and I wanted my own. So I began bringing back a little bit of water from every place I went—some for Luisah and some for me.

At that year's ritual for Brigid, the Irish Goddess of the holy well and the sacred flame, we decided to create a holy well out of a punch bowl. I added all my waters, and others brought sacred water of their own, or simply brought the water they drink every day. We made a pledge to Brigid, and saved back some of that water to seed the next year's ritual.

Over the years, the tradition grew. I began carrying the waters of the world around with me in a small bottle, using it to make offerings to the land or to the waters I was sampling. We began asking people to bring water at the start of many rituals, not just Brigid but Witch camps and political actions and other gatherings. As I describe at the start of this book, we make an offering of water to begin our fire ritual every year.

People began sending us water, going to special places to collect it. Some of our waters of the world came from Ireland, from many of Brigid's wells. And we got waters from sacred rivers—from the Ganges and the Nile—from Chalice Well in Glastonbury, from the *pozo* in the town of Amatlan, where Quetzalcoatl was born, from every continent. Someone sent us Arctic waters, and someone else a bottle of melted ice from Antarctica.

A few drops of waters of the world turn any vessel of water into sacred water. It is endlessly replenishable, not subject to scarcity. We use it to honor the spirits of the water and the land. One taboo: we don't drink it. "Sacred" is not necessarily the same as "sanitary," and too many of the world's waters are, regrettably, polluted.

HONORING THE WATERS OF THE WORLD

To honor the waters of the world in ritual, we set a vessel of water in the center of the circle. One by one, people come up, add their water, and say where it is from. Then a priestess lifts the bowl and says whatever is in her heart. Generally I say something like

> *Spirits of the waters, spirits of the land, ancestors, spirits of this place, we bring you this water as a gift of gratitude, for this land, for letting us stand and walk and be here. We bring it as a sign of respect, a sign that we want to open our ears and listen to what the land has to say to us, that we*

want to learn to be healers and good allies of this land and its people. Water is life, and this water comes from many places, as does the blood that flows in our veins. We ask your permission to stand in this place, to root here, to move energy here and do our rituals here. We ask your help in opening to what this place has to teach us. We ask your help that someday all the waters of the world may be clean and run free. We thank you for the power and beauty of this land, and for the gift of life.

The priestess then sprinkles water in the four directions, and also above, below, and center. The water remains in the circle throughout the ritual, and when the ritual is done, some is kept back to add to the waters of the world, and the rest is poured out onto the land.

OBSERVING WATER

Observing water is a meditation in itself. Just watching a flowing river or a running stream can help us feel calm and renewed. Swimming, floating, being in or near water is one of the basic ways human beings relax and replenish our energies.

Following are three suggestions for observing water.

Observing Water's Effect

In your home base spot, ground and come into your senses. Look around at the form of the land, the plants, the shape of the hills, the creases and crevices. Become aware of the presence and traces of water, of the flows that have shaped the land, smoothed the rocks, of the water that permeates the soil, the water encompassed in the bodies of plants and animals. Observe the presence and flow and movement of water.

Observing Water in Motion

Sit beside a running stream, or a swift river, or the ocean. Watch the movement and form of the water. Notice the shapes and patterns that it makes, where it runs fast and where it slows down, where there are standing waves and where there are slow eddies. Notice the way the patterns of movement form and reflect the shapes of the land. The visible motion of water is only the surface layer of more complex movement below. What can the surface tell you about the depths?

Immersion

Get into water. Go swimming in a river or bodysurfing in the ocean. Be sure to be safe, have a buddy, and be aware of currents and undertows. Feel the force of the water on your body. Notice how you move in the water, how the waves and ripples feel. Dive down and feel the difference between the motion below and the motion above. Feel the temperature changes from the depths to the surface. Close your eyes, and observe the water with your skin, your muscles, your deep bodily senses.

Water Cycles

To come into relationship with water, we must understand how water works. We must treat it respectfully in very practical ways, and learn its cycles, in order to hear its deeper communication.

Life began in water, and water remains necessary to life. Plants may use the sun's energy for photosynthesis, but it is water and carbon dioxide that they break down and recombine into the carbohydrates of sugar and starch. And while the sun's energy is virtually unlimited, in many places in the world water is the limiting factor.

My garden is dependent on that small pool in the spring box. Our land receives more water than most places on earth—eighty to a hundred inches in a good year. But almost all that rain comes between November and May. We also have one of the longest dry seasons in the world—four to five months a year when essentially no rain falls at all. We are a land of extremes.

Because of those extremes, when I plant my garden I have to plant for the water I'll have available in July and August, not for the abundance I can expect in January. I have to consider how to store and conserve water, how to make the most of that little trickle flowing into my tank.

But where does January's abundance go? Each winter this land is drenched by rain, the equivalent of a lake five to eight feet deep covering every square inch of these hills. By midsummer, streams have dried to a trickle at best, bone dry at worst.

Some of the water, maybe 80 percent, runs off the land, in streams and creeks and rivers.

Some of it soaks into the soil and remains there, coating the particles and filling the spaces within the soil's structure, or sinks deep beneath the earth to pool when it hits an impervious layer of soil, either gathering in an aquifer or forming a spring and making its way to the surface again.

Some of it is in these trees towering hundreds of feet above me. A big redwood can cycle seven hundred gallons of water a day, more than the output of my little spring.

The redwoods drink water from the ground, but they also comb fog from the air. Their needles condense mist into droplets, creating their own "rain." We get no true rain in the summer, but the inland heat often pulls in fog from the coast. Redwoods can live only in fog zones. The big ones are actually too tall to pump water from below all the way to the top, and their crowns are sustained by the moisture they pull from the air. In the old growth, a whole world of fungi and lichens and other plants grows high in the canopy. Over a thousand different species live high above the earth, sustained by moisture in the atmosphere. Some of them don't even begin to grow until a redwood is 150 years old.

Where does the water come from? Gaia is more blue than green. There's an enormous amount of water on the planet, the vast majority of it in the oceans or locked in the ice of the poles. The rotation of the earth keeps the oceans stirred up and cycling, swirling in great currents that moderate the climate of the world and cycle nutrients around the seas. The pull of the moon's gravity shifts the seas back and forth in the great tides that swell and circle the planet twice each day. The sun's heat pulls water up from the depths, to evaporate and ride the skies as the vapor that forms the clouds. When the clouds from the ocean cool and touch these hills, they drop their rain. The rain falls, sinks into the earth, feeds the roots of these trees, or runs off into streams and rivers that eventually find their way back to sea. The cycle is complete.

Water has always been symbolically linked to our emotions. Our feelings ebb and flow, storm and subside, just as water does. Thus water can teach us something about emotion.

WATER TRANCE

Begin in a safe place where you will not be interrupted. You might choose to lie down to experience this trance, or to stand up and dance and move with it. If you are doing the trance in a group, one person can read the trance instructions for others. Better yet, that person can become familiar enough with the journey to make up her own words.

We often begin by singing and dancing ourselves into a light trance state, using the following chant:

> Born of water,
> Cleansing, powerful,

Healing, changing,
We are.

Breathe deep. As you dance and move, feel the water in your own body. We are made up of mostly water, and our bodies feel the pull of the tides and the rhythms of the moon. Our blood is seawater in a new form, and we can still feel within us the crash of the waves and the great, slow currents circling the earth.

And as you feel the water within you, gradually let your skin and bones and human self drop away. Feel yourself drifting and floating, calm and peaceful, until you become a single drop of water, floating in a pool deep beneath the earth.

Let yourself drift for awhile, feeling how calm and peaceful and still it is here, in the dark, with everything that is not you stripped away. Feel the power of water to be pure essence. Breathe deep, and take in that power.

But even in this stillness, something moves. Time passes. High up above you, the sun calls. Something shifts, and you begin to rise.

Higher and higher you rise, finding the cracks and crevices in the earth, until at last you emerge, under the sun, as a fresh spring of water.

Feel the joy of the sun on your face and the bright sky above you. Notice what comes to drink from you, what lives on your edge. Is there some place in your life where you need this power of water to emerge, to renew? Breathe deep and take it in.

And you are so filled with joy that you spill over and begin to flow. The spring becomes a stream, bubbling and dancing down the hillside, noticing how it feels to leap over rocks and sing. Is there a place in your life where you need that power of water to dance and play? Breathe deep, and take it in as you sing,

I am the laughing one.
I am the dancing one.
I whisper secrets
As I flow.

And you flow on, growing stronger as new streams join you. One stream merges with another. Think about where in your life you need that power to join with others, to merge. Breathe deep and take it in.

You grow stronger and deeper, carving the hillsides as you pass, smoothing the boulders, tumbling the rocks. Feel the power of water to shape and change what it touches. Is there a place in your life where you carve your own channel, where simply by being who and what you are you create change? Is there a place where you need that power? Breathe deep, take it in, and sing it out:

I am the shaper.
I am the changer.
I carve the mountains
As I flow.

And now you come to something that blocks your flow. Take a deep breath, and see and feel what this obstacle is. Does it remind you of anything in your life? Is there something written on it? Does it speak to you?

Water has many ways to move around obstacles. It can crash through them and wash them out. It can dissolve them slowly. It can move around or under them. It can back up behind them until it flows over them.

How do you move beyond this obstacle? Take a deep breath and think about all the powers of water and the power you have within you. Let your breath become a sound of power, a sound you can share and blend with others, a sound that can carry you past this block.

(Wait a moment while the sound dies down.)

Now breathe deep and look behind you. What has shifted? What has changed? Where are you now, and what power do you feel within you?

And you flow on, down through the hills and mountains, out onto the wide valleys, growing stronger and deeper. And now you spread out and flow as a great river, moving through the valley toward the sea. Feel what lives in you, what you carry, what grows along your banks. Know the power of water to nurture and sustain. Is there a place in your life where you need that power? Breathe deep and take it in.

As you flow, notice how your back grows warm from the sun, how the warm water rises and then is pushed aside by cold water warming from below, until you begin to turn like a spiral cycling up and down and around. That cycling motion becomes a pulse, a meander. You carve the banks of the river and drop silt in the bends. You curve and snake, refusing to flow in a straight line. Feel the power of water to meander, to flow in its own shape. Is there a place in your life that meanders, that does not move swiftly from one goal to the next but takes its time, finds it own route? Do you need that power? Breathe deep and take it in.

The river is flowing,
Flowing and growing,
The river is flowing
Down to the sea.
Mother, carry me,
Child I will always be,

Mother, carry me
Down to the sea.

And at last you reach the sea. You branch out into an estuary and seep out into the waves. Feel the power of water to spill itself out, to become something larger than itself. Is there a place in your life where you need that power? Breathe deep and take it in. Sing it out.

The ocean is the beginning of the earth.
The ocean is the beginning of the earth.
All life comes from the sea.
All life comes from the sea.

Feel the wild wind on the waves, as they roar and crash and beat down the shore. Breathe deep, and feel the power of water to rage, to tear down. Is there a place in your life where you need that power? Breathe it in.

And now feel the ocean grow calm and still. The wind dies down, and the waves lap quietly against the shore, mirroring the sky. Is there a place in your life where you need that power of water to soothe, to be tranquil? Breathe deep and take it in.

Sink down, deep down, below the realm of sunlight, down to the bottom. Feel the ocean's incredible depths. Dark, cold, mysterious, pressed down by the weight of all the water and life above, they nevertheless can be a source of the nutrients needed for life to thrive, a creative source. Feel your own depths, the creativity that surges upward from deep below the light of conscious awareness. Take in the depth and power of water.

And now feel the sun pulling you upward again. Rise, up toward the light, until you lie on the surface of the waves, dancing with the wind, getting lighter and lighter. Is there a place in your life where you need that power of water to lighten up? Breathe deep and take it in.

And eventually you become so light that you evaporate. You become vapor, a gas on the wind, flying high above the earth, dancing in the clouds.

And you look down below, and you see the earth beneath you. You see places that are green and flourishing, and places that are barren. You see places that are whole and thriving, and places that are wounded and healing.

And eventually you see a place that calls to you, that needs the nourishment and life that only you can bring.

You feel cooler, suddenly—heavier. Your place is calling. And you congeal, back into a drop of water, and fall.

You fall and fall and fall, until at last you reach the ground and sink in as a drop of rain. And you know the power of water to give itself away and let go.

And you sink into the earth, feeling the roots that drink from you, the tiny creatures that swim in you. But you continue to sink, down and down, through sand and soil and gravel, until anything that is not you is stripped away.

And at last you become simply a drop of water, floating in a deep pool beneath the earth, knowing what it is to be only the purest essence of what you are.

Now thank the water, for its teaching and its power and its journey. Feel how that cycle of water, with all the power it embodies, lives in you.

But begin to remember, now, that you are a human being, that while you are mostly water, you are not only water. Feel your bones begin to grow back, your skin again containing the water within you. Move your awareness back into your human mind, thanking the water and remembering its teachings.

Let's sing ourselves back:

> Born of water,
> Cleansing, powerful,
> Healing, changing,
> We are.

If you've been lying down, slowly sit up. Feel the edges of your skin, the solidity of your bones. Say your name, and clap your hands three times.

And that is the story of water.[1]

Water and Abundance

From my journal:

Mer and Ken dug out the spring yesterday, while I stayed home to write. Three of us can't actually work up there at the same time anyway—there's not enough room. The good news is, after they removed masses of roots that had grown into the silt and gravel around the spring box, they got down to the true flow, which is abundant, and now we're getting four times as much water as we were before! We can water the gardens. I can take a slightly longer shower. We might even be able to refill the pond in the greenhouse!

I feel rich. I feel as if someone just told me I had four times as much money in the bank as I thought. In fact, somehow the state of my water tank and the state of my finances have become psychologically linked for me. I guess both ultimately lead to food. Both require sources, reserves, and outflows, and both are capable of springing unexpected and disastrous leaks.

We still don't have water to squander, but we do have enough to meet our needs, and some left over. And that's abundance!

Water creates abundance. Water makes it possible for plants to use the sun's energy to create food. Abundance comes not just from how much water flows into a place, but how well it is used, whether it is available where and when it is needed, and how many times it can be recycled and reused before it flows away. I can create abundance by increasing my sources of water, as in the journal entry. I can increase my reserves, or store more water to last through the dry summers. Or I can work out ways to reuse the water. In fact, we've done all these things: built ponds and cisterns, added water tanks, and installed graywater systems (more on those later).

The best place to store water is in the soil. Here in northern California, the earth does that for us, at least in part, soaking up those heavy winter rains and releasing the water that percolates down through the ground as springs. Our springs are all "perched springs," meaning that they rest on an impermeable layer of clay and are fed purely by each winter's rains. Few people drill wells here—a hole poked through that impermeable layer can drain your neighbors' springs as well as your own, and there are no underground aquifers to tap.

In other areas, however, massive lakes of water exist deep underground, amassed through centuries. They can be tapped by deep wells, providing abundant water. But if they are tapped more quickly than they can be replenished by each year's rains, sooner or later they will be drained dry, just as my water tank will empty if I use more than comes in each day.

We are currently using up our stores of water at a phenomenal rate. In California, Owens Lake, once the third largest body of water in the state, has been sucked completely dry to feed thirsty Los Angeles.[2] Groundwater is being pumped faster than it can be recharged almost everywhere that humans have discovered it. The Oglala Aquifer, one of the world's largest underground water resources (which supplies the dry states between Texas and South Dakota), is already 60 percent tapped out after only a few decades of water mining.[3] Water demand in that region tripled between 1950 and 1990, and is expected to double again by 2015.[4]

Maude Barlowe, national chair of the environmental justice group Council of Canadians, writes,

> Global consumption of water is doubling every 20 years, more than twice the rate of human population growth. According to the United Nations, more than one billion people on earth already lack access to fresh drinking water. If

current trends persist, by 2025 the demand for fresh water is expected to rise to 56 percent more than the amount that is currently available.[5]

But we don't have to be draining our water reserves. There are many ways that even city dwellers can increase our abundance.

Let's consider storage first. As I've said, the best place to store water is in the soil. Living soil is like a sponge. It's porous, full of spaces that can hold water and air. Soil rich in organic matter can hold many times its weight in water.

On a small scale—say, your garden—the first way to encourage water storage is to make sure that the water you give your plants can't easily evaporate. There are two ways to do that. The first is by closely spacing your plantings, which creates a canopy of green over the garden bed. This requires a fairly high level of fertility in your soil, but has the added advantage of producing a lot in a fairly small space.

The second evaporation-discouraging method is mulch. Mulch is basically stuff you throw on the ground to cover it—ideally, made of organic materials. Mulch can be straw, dead leaves, dried grass clippings, last season's corn stalks or tomato vines, etc. Pile it on the garden, as high as you can, and plant through it or heap it around your plants, being careful not to smother their crowns.

A heavy cover of mulch has a number of advantages. Not only does it keep water from evaporating; it also softens the impact of splashing water drops, whether from rain or from that high-pressure jet your six-year-old lets loose from the hose. Mulch also feeds the life of the soil. It nurtures worms and soil bacteria, those creatures who contribute so much to fertility. Get enough worms tilling your soil, and mulch can save you the backbreaking work of digging, turning, and fertilizing your garden. By adding organic matter to the soil, you increase the soil's ability to hold water as well. You can set in motion a very beneficial self-reinforcing cycle: mulch decreases evaporation, increases the water held in the soil, and improves fertility, so your plants grow more lush and abundant; this in turn gives you flowers, fruit, and vegetables that give you not only food but also stems and stalks for mulch.

Water can also be stored in the soil by contouring the ground. On a larger scale, permaculturists dig swales—ditches with berms that run along the contour of the land—to catch running water and hold it so that it infiltrates the ground. Swales can be dug by hand (with a spade and hoe) or, if larger swales are desired, with earth-moving machinery. As water infiltrates over time, the swales build up a "lens" or micro-aquifer of stored water that can be tapped by trees, shrubs, and other deep-rooted plants. In the long run, swales often fill in and become terraces. In many dry climates, hillsides have been terraced over the centuries to catch and conserve water, changing the face of the hills themselves. In Spain, Italy, Greece, Nepal, and throughout the Middle East, sculptured hillsides proclaim the antiquity of agriculture.

If your garden is on sloping land, a swale or two might give you the outline for terraces and garden beds. Put the beds on the downside of the swale, to take advantage of the water collected, and plant drought-tolerant herbs on the upside. If your garden is flat, small mini-swales and channels can help more water infiltrate. Dig low paths between beds, and mulch the paths, and they will serve as infiltrators. (Don't, however, mulch them with slippery straw. Try dead leaves or wood chips.)

Water can be stored in ponds, too. The technical aspects of pond-building are beyond the scope of this book, but advice and instructions are easily available. A small backyard pond can easily be built in an afternoon, and will add much to the life of your garden.

Every garden should have a pond. Even a small pond can provide habitat for beneficial insects and a whole variety of pest-eaters: frogs, toads, birds, even turtles and snakes. A few minnows will keep mosquitoes from breeding. Water plants are some of the most efficient users of the sun's energy, and a pond can be a great source of nitrogen-rich materials for compost and mulch. You can grow food in your pond—water chestnuts, even fish if it's big enough. Plus it gives you a good way to use all those sacred rocks you've been collecting!

For children, a back-yard pond can be an introduction to the pleasures of watching nature. Seed it with frog eggs in the spring, and later watch the tadpoles grow into frogs. Inoculate it with water and muck from a natural pond, and see what grows. Many of the city-raised kids I know spend most of their free time inside, staring at computers or homework or TV screens. To entice them outside, we need to provide something equally fascinating, and a living pond can catch and hold their interest.

Don't ever leave very young children unsupervised near an unfenced pond, of course. Don't dam a running stream to make a pond or place one in a live waterway without long consideration and expert help, or you risk causing terrible erosion when storms occur. Always give a pond an outflow. And never put non-native water plants into a pond linked to an ecosystem's waterways. In our area, water hyacinths don't overwinter and my ponds don't link to running streams, but in many areas, water hyacinths and other water plants have become a menace. Near where I live, our area's reservoir (and main swimming hole) has become choked with elodea from somebody's dumped-out fish tank.

And, finally, water can be stored in tanks, cisterns, even simple rain barrels filled from roof runoff. We don't all have a spring, but if we're lucky we do have a roof, and that roof by its very nature intercepts the rain on its way down.

Instead of channeling that water into gutters that drain into the sewer system, sometimes overloading it in heavy storms, we could catch the water from our roofs and use it to water our gardens.

Roof catchment is the major source of water for many people who live on islands, or land where water is scarce. If you live in a climate that gets intermittent rain, catching and storing even some of your roof water might provide all the water you need for a flourishing garden, reducing your water bill and saving some of Gaia's squandered gift.

How much water are we talking about? Let's say your roof covers a modest home of 1,200 square feet, and you get 24 inches of rain a year. That's 2,400 cubic feet of water each year. A cubic foot of water is around 7.5 gallons. Taking that number times the cubic footage, that's 18,000 gallons of water a year. A typical garden of 1,000 square feet can thrive on a generous 100 gallons a day, so that much water could keep your garden green for 180 days![6]

Of course, storing that much water above ground would need an enormous amount of space: two cisterns twenty feet square and thirty feet high, for example. Few of us have that much space to spare. But in most places, rain comes intermittently enough that we don't need to store a whole year's worth of water. Even a couple of fifty-five-gallon drums at your downspouts could provide you with some extra water for emergencies, or with soft rainwater for your hair or handwashing or ritual blessings.

And much of that water can be stored in the soil. My friend Erik channels all of his roof water into the ponds and swales of his suburban backyard, building the water lens, that underground store of water below his garden, to help his plants through bad times.

We can also increase our water abundance by reusing water. The water from a shower or bath or from cleaning vegetables in your sink could be growing your food instead of running down the drain into the sewage system.

Such water is called graywater—not as dangerous to treat as raw sewage, but still not entirely safe to dump on your lettuces. Graywater carries bacteria from your skin, soil bacteria from the roots of those vegetables, grease, oil, soap residues, and whatever else gets dumped down the drain. It can be a fertile medium for bacteria and other beasties to breed in. But graywater is also easy to treat.

I remember, during my first permaculture class, when my friend Penny Livingston-Stark announced that we were going to learn how to clean water. I got very excited, and then mad. How to clean water—doesn't that seem like something everybody should know? And why, in something like eighteen years of formal education, had nobody ever taught me?

Living organisms clean water. In *Water: A Natural History,* Alice Outwater describes the many ways water was once kept clean and clear by plants and animals. She explains the roles beavers, prairie dogs, wallowing buffalo, and mussels once played in collecting, storing, infiltrating, and cleaning our waters, before trapping, hunting, plowing, and engineering destroyed these natural systems.[7]

Swamp plants remove excess nitrates from water. The bacteria that live around their roots gobble up fecal coliforms and other nasty beasties. Some species also take up minerals and even heavy metals. Constructed wetlands are capable of cleaning sewage. The city of Arcata, for example, has constructed a marsh that not only handles the city's sewage but provides its largest tourist attraction as habitat for many birds and wild animals. John and Nancy Jack Todd, of the Ocean Arks Institute, in Massachusetts, have pioneered the development of "living machines"—water-cleaning systems that work on the same principle as the wetlands, running sewage or contaminated water through a series of tanks containing different biological communities, from algae up to canna lilies and fish. Clean, drinkable water comes out the other end, at a fraction of the cost of conventional chemical treatment systems.[8]

A home graywater system can reuse water fairly simply. The very simplest might be to place a bucket with you in the shower, to catch your warm-up water and the splash from rinsing yourself, and then use that on ornamentals, not edible plants, in the garden or to flush the toilet. The next step up would be to hook your sink or washing machine into some kind of "biofilter": a holding tank filled with gravel, lava rock, or some other medium where helpful bacteria can grow. The tank can be covered, to prevent mosquitoes breeding, or left open as long as the water stays below the surface level. An open tank can grow water plants whose roots will also help the breakdown of dangerous bacteria. Water plants infuse oxygen into the water through their roots, creating a zone in which aerobic, or oxygen-breathing bacteria, can live. A tub containing gravel and water plants will provide habitat for both aerobic and anaerobic beneficial bacteria, which can chomp happily away on fecal coliforms and other potential disease carriers.

The water can then run from the tank into a sequence of tanks, or into a gravel bed, a small constructed wetland in your backyard, a pond, or a gravel leachfield under your plantings.

Any system that works with living things will require monitoring and adjustment. Penny runs her graywater into a duckpond. In its first incarnation, the ducks refused to enter the water. Surfactants, the agents in soap that break down oils, were still present and would have ruined the duck's feathers. Penny added another biofilter, ran the water through a small artificial stream to aerate it, and then through a gravel bed, and now the ducks enjoy their pond.[9]

Maybe you won't put down this book and go redo your gutters or put in a graywater system. Perhaps you live in an apartment and have no garden to water. But we can all bring ourselves into right relationship with water—if nothing else, by conserving it. We can turn off the tap while we brush our teeth or do the dishes, save our warm-up water from the shower, take showers instead of baths, and make them shorter showers or install a low-flow shower head. We can be conscious of water, express our gratitude when we do use it, avoid disposing toxic substances into it, and treat it with respect, as the sacred gift that it is.

Water Policy

The home-scale solutions I've outlined above are good models for the potential solutions we could put into place on a larger scale, by changing our water policies. As Witches, as people who believe that water is sacred, we should be advocating for large-scale changes in the way we treat this precious substance.

When so many elegant, cost-effective solutions exist, why aren't we putting them in place? Why isn't it the norm to build a roof-catchment/storage/irrigation system into every unit of new housing? Why are we still building costly, inefficient, environmentally damaging sewage systems?

There are two basic reasons. The first is that new ideas always meet resistance, and changing our way of thinking about things is sometimes harder than changing the things themselves. It's so much easier to stick with the tried and true than to risk the unknown unknowns.

The second reason is that many people have enormous interests vested in the system as it is. Construction, sewage treatment, water provision—all these are areas where enormous profits can be and are being made, and those who profit wield enormous influence over legislators, bureaucracies, and enforcers.

What can we do, as individuals? We can begin by educating ourselves and our communities, learning how water works and what the solutions to our problems are. We can try out solutions on a small scale to fine-tune them and show others how they work. We can ignore the rules that maintain the status quo and do things anyway. We can strongly oppose bad policies.

Here in northern California, our larger community defeated a scheme by a corporation to take water from the mouths of the Gualala and Albion Rivers and tug it down to San Diego in giant plastic bags. Besides the almost ludicrous nature of the proposal, and its potential impact on the ecology of our river systems, we were alarmed to discover that the corporation proposing it was strongly influential in the World Trade Organization. Under the rules of many international trade agreements, such as NAFTA (the North American Free Trade Agreement), corporations can sue governments for loss of projected profits if the governments pass laws interfering with their business operations, even if those laws are for public health, safety, or other good! Had the waterbag scheme succeeded, it would have meant that all of northern California's waters were now opened to profit-making, and laws regulating the sale and privatization of water would have been very difficult to make or enforce.

Electoral politics can be a fruitful field for intervening in issues around water. Many crucial decisions are made by water boards, who are often elected with very little opposition or competition. Relatively small investments of time or money can yield a high degree of influence over local directions. Rightwing fundamentalists have gained an enormous amount of political power by running for local school boards. Why shouldn't people who care about the earth gain power by running for water and utility boards?

Water and Scarcity

Water, it is predicted, will be the great issue of the twenty-first century, the center of resource wars and conflicts. Because we haven't yet been courageous enough to implement sane solutions, and because control of the world's water is becoming more and more concentrated while the population is growing, it is estimated that by the year 2020 two-thirds of the world will be without adequate supplies of clean water. Water has always been seen as a communal resource, something that should belong to all, and water delivery has long been a primary function of government, something we pay for and make decisions about *collectively*. Today, there is more and more pressure to privatize water, to place its ownership and control in the hands of corporations that can make a profit out of providing this basic human need.

What does water privatization mean? How many of us already filter our water, or buy bottled water to drink? When I was growing up, we assumed our tap water was drinkable—that was one of the hallmarks of development and democracy that America supposedly stood for. Today, we trust that our tap water probably won't give us typhoid or cholera, as it might in Mexico or India, but we suspect that it might give us cancer. Privatized water services in England, France, and Wales have meant increased rates and lack of access to water for many low-income users. In Bolivia, water privatization resulted in a 40 percent increase in cost and sparked an uprising in Cochabamba in 2000.

Maude Barlowe writes,

> Already, corporations have started to sue governments in order to gain access to domestic water sources. For example, Sun Belt, a California company, is suing the government of Canada under NAFTA because British Columbia banned water exports several years ago. The company claims that B.C.'s law violates several NAFTA-based investor rights and therefore is claiming US$10 billion in compensation for lost profits.[10]

Among the world's poorest people, who can least afford to pay for the basic necessities of life, water privatization is well advanced, often imposed on third world countries by the International Monetary Fund or by provisions of global or regional trade agreements. Most often, this means higher prices and reduced services. In Cochabamba, Bolivia, for example, the city's water supply was privatized in 2000, its control given to a company called Aguas de Tunis, a subsidiary of Bechtel. All sources of water, even those privately held, were covered. (If those provisions applied in my area, Bechtel could charge me for the water I take from my own spring!) Water prices tripled, and many people were paying a third or more of their income for water.

The people of Cochabamba rebelled. They staged a nonviolent uprising, blocking roads and commerce in the city for two weeks in April of 2000. The government eventually gave in and turned water delivery over to a committee of the people, called La Coordinadora.

Oscar Olivera, one of the leaders of the uprising, told a group of us when he visited San Francisco in 2002 that the organizers referred so often to La Coordinadora during the conflict that many people thought they were talking about a woman (since the term is feminine). Who was this larger-than-life woman, they wondered, who would provide water, distribute it equitably, and take care of their needs? Perhaps, without knowing it, they had invoked a new aspect of the Goddess.

The people of Cochabamba wrote the following declaration.

Cochabamba Declaration on the Right to Water

> Here, in this city which has been an inspiration to the world for its retaking of that right through civil action, courage and sacrifice standing as heroes and heroines against corporate, institutional and governmental abuse, and trade

agreements which destroy that right, in use of our freedom and dignity, we declare the following:

For the right to life, for the respect of nature and the uses and traditions of our ancestors and our peoples, for all time the following shall be declared as inviolable rights with regard to the uses of water given us by the earth:

1. *Water* belongs to the earth and all species and is sacred to life, therefore, the world's water must be conserved, reclaimed and protected for all future generations and its natural patterns respected.

2. *Water* is a fundamental human right and a public trust to be guarded by all levels of government, therefore, it should not be commodified, privatized or traded for commercial purposes. These rights must be enshrined at all levels of government. In particular, an international treaty must ensure these principles are noncontrovertible.

3. *Water* is best protected by local communities and citizens who must be respected as equal partners with governments in the protection and regulation of water. Peoples of the earth are the only vehicle to promote earth democracy and save water.[11]

The Living River

Inspired by the people of Cochabamba, our network of Pagan activists has participated in many actions around water, privatization, and corporate control of the environment by creating what we call the Living River.

The Living River began with the Quebec City demonstrations against the summit meeting of the Free Trade Area of the Americas (FTAA), the extension of NAFTA throughout the western hemisphere, in April of 2001. We formed a "river" of participants dressed in blue, carrying flowing blue cloth and following a giant blue river Goddess puppet. Our goal was to bring the Cochabamba declaration into the meetings, to say, "This is what we should be negotiating: the right to water and the need to preserve it, not the opportunity to privatize and profit from it." A nine-foot-high fence, several thousand riot police, and a barrage of tear gas kept us from entering the meetings, but we read the declaration at the gates and were able to focus attention on water issues. Since then, we have carried on this tradition at many demonstrations. We took the Living River, together with Oscar Olivera himself, to the doors of Bechtel

Corporation, one of the world's largest water privatizers, which is currently suing Bolivia for $40 million in loss of projected profits for declining its efforts to profit from their water.

The Well of Grief

Water is emotion, and water is also key to cleansing and healing. When we begin to open up to the natural world, when we drop our defenses and begin to hear what nature is saying to us, when we start to appreciate the incredible beauty and wonder of the world, we also become aware of how much is being destroyed. Grief and sadness may overwhelm us at such times.

Jon Young, director of the Wilderness Awareness School, talks about how difficult it is to train young people as trackers. The skills and techniques, even the stillness and consciousness, are not hard to learn. But when people open their awareness past a certain point, they hit what Jon calls the "wall of grief," an experience of being overwhelmed with sorrow at the loss and degradation of the natural world around us.

I experience it more as a *well* of grief, an upwelling of sorrow and tears that seems to come from the very heart of the earth, dark and cold and unimaginably deep.

Grief is like water: we can drown in it, but we can also drink from it and be strengthened and nourished by it. For our grief and sadness reflect our ability to feel, to love, and to mourn the loss of what we love so dearly.

One of the great gifts we can give each other is simply to listen to the grief, pain, anguish, fear, or whatever emotion someone needs to express, without trying to fix it, change it, or take control of it in any way. If someone is suffering from trauma or post-traumatic stress, she needs to tell her story, sometimes over and over again. When we are in grief, we need to share it with someone who will not himself be overwhelmed or shocked by it.

LISTENING MEDITATION

In pairs, decide which partner will speak first. Ground, and then just breathe together for a moment, matching breaths.

The first partner speaks first, telling about an intense experience. It can be a sad or a happy or a frightening experience—whatever that person feels moved to speak about. The second partner just listens, trying to listen on every level—to the words and content, to the energies, to the emotion, to the body language. Listen for at least three minutes without interrupting, asking questions, or offering comments. Then switch roles.

Afterward, talk about how it felt to listen so intensely, to be listened to so well. What happened to the energies between you? Within you? In the room? How do you feel about each other now? About the experience you talked about?

Grief Ritual

Ritual is one of the tools that can help us with grief. The key to a powerful ritual for grief is to open a space in which people can speak from their hearts, can name what they may not have been able to speak about before, with energetic support. In Chapter Seven, we learned about silent cheering. Energetic support in a grief ritual may not be so cheerful: it might come in the form of chanting, holding a low tone, or simply witnessing with deep attention.

Two summers ago, I was teaching in the British Columbia Witch Camp, together with Sunray and Culebra. We created a simple but beautiful healing ritual.

We placed a bowl of waters of the world in the center of the circle, after grounding, casting, and doing all the preliminaries. We also had a bowl of pebbles, and each person in the circle was given one.

One by one, each of us stepped forward and dropped our pebbles into the water, naming what we grieved for or what needed healing.

We sang a healing chant that spoke of going down to the water and letting our tears and fears be washed away. Singing, we carried the cauldron, now heavy with stones, all the way down through the camp, down the mountainside, and down to the lake, where we stripped off our clothes and jumped into the cold, cold water. We carried the cauldron in, and two women dumped the heavy load of stones into the water. The chant became joyful, jubilant as we felt cleansed and alive, burning with cold. Then, still singing, we marched naked back up the hill, completing the ritual by raising power, that is, by letting the chant build until its energy reached a peak of release. In the quiet that followed, I felt myself able to let go of some of the pain I was carrying from seeing people brutally beaten by police in the Genoa demonstrations against the G8 summit earlier that summer. The grief and sorrow remained, but my own energy was flowing again, no longer stuck and inward-turning.

I lead many grief rituals, for activists, and for all of us suffering the pain and loss of ordinary life in these difficult times. A communal ritual for grief can allow us to acknowledge and share deeply some of the emotions we ordinarily keep hidden. Together we can support each other through the sorrow, and come closer in the process. The love and compassion we share is the true healing.

A grief ritual should not try to artificially stir up emotions or push people to emote. The most powerful rituals happen when we simply create an opening and an atmosphere of receptivity, as in the ritual described above. I often use water in some form to symbolize this receptive state. Waters of the world are called for in the above ritual, but salt water or clear spring water would work just as well at the center of the circle. Likewise, we used small stones in the example above, but salt or any other object could be used as a symbol of release.

Grief may also be linked to rage. Sticks and something that can safely be hit, such as a big plastic barrel, can also be placed in the center.

The ritual also needs some way to allow us to use our voices and express our sorrow. Keening—the traditional Irish form of lamenting the dead—crying, even screaming, work sometimes, but people often find it hard and artificial to keen on command. Singing can open the emotions and provide a base of sound that allows people to moan, cry, or sob if they feel moved to do so under the cover of many raised voices. And music has its own healing power.

When the energy is released in a grief ritual, be sure to do something to cleanse, renew, and fill up again afterward. We bathed in the lake in the ritual example given here. At other times, I have plunged into the ocean, washed my face in a bowl of salt water, spilled the water of grieving onto the ground, or shared food and drink to symbolize nurturing. Feel free to create the ritual that will best serve your needs and community.

Grief, trauma, and fear can lead us to close down. When we shut down to negative feelings, we often shut off our ability to feel positive emotions as well, to take pleasure and joy in what remains of nature, to give and receive love. We may become angry toward our friends and lovers, cynical and bitter, difficult to be around or to live with. In extreme cases, grief and trauma can lead to true post-traumatic stress syndrome, deep depression, and even suicide.

But when we find someone willing to listen, when we find the support of a community that can hold our grief and not turn away, we can find the courage to open up again. Then we can drink from the well of *life*, not just the well of *grief*, and be healed.

BLESSING FOR WATER

Praise and gratitude to the sacred waters of the world, to the oceans, the mother of life, the womb of the plant life that freshens our air with oxygen, the brew that is stirred by sunlight and the moon's gravity into the great currents and tides that move across the earth, circulating the means of life, bringing warmth to the frozen Arctic and cool, fresh winds to the tropics. We give thanks for the blessed clouds and the rain that brings the gift of life to the land, that eases the thirst of roots, that grows the trees and sustains life even in the dry desert. We give thanks for the springs that bring life-giving water up from the ground, for the small streams and creeks, for the mighty rivers. We praise the beauty of water, the sparkle of the sunlight on a blue lake, the shimmer of moonlight on the ocean's waves, the white spray of the waterfall. We take delight in the sweet singing of the dancing stream and the roar of the river in flood.

We ask help to know within ourselves all the powers of water: to wear down and to build up, to ebb and to flow, to nurture and to destroy, to merge and to separate. We know that water has great powers of healing and cleansing, and we also know that water is vulnerable to contamination and pollution. We ask help in our work as healers, in our efforts to ensure that the waters of the world run clean and run free, that all the earth's children have the water they need to sustain abundance of life. Blessed be the water.

TEN

Earth

From my journal:

Sometimes the better part of gardening is not doing it. Today my friend Caerleon and I went for a walk. Originally it was planned as a short get-in-shape walk—but the day was perfect, a sunny winter day with the hills green and the sky blue, the sun warm but not too warm. We walked down our dirt road to the gate of the neighboring ranch and then for some reason continued on, down to the stream that forms the headwaters of the Gualala River, and back up an enticing dirt road that had always intrigued me.

This whole area is carved by the logging roads put in fifty years ago. Haunted by the ghosts of great trees that stood here before the loggers came, they determine the routes we travel, the paths of erosion. At best, they make great hiking trails. We climbed up beside the stream bank, looking down an almost vertical hill to see the enormous stumps of huge Douglas firs that are no more. The road led us to a meadow surrounded by a bowl of green hills, and then zigzagged up a steep grade to join the dirt road above the Big Barn, left over from the days when this land was a sheep ranch. We walked down, across a broad, green meadow, speculating on what the land would have looked like before it was logged. Would these open fields have been covered with trees, or were they natural meadows? Why haven't the trees come back if once they covered these fields? Grazing was the culprit, we speculated—

especially here, so close to the barn that was once the hub of the old sheep ranch. I explained about the mycorrhizal fungi, while Caerleon, who is an archaeologist, told tales of the ancient tribes that once lived here.

Back down by the stream that runs alongside the road, we climbed a blue schist boulder, a pitted rock marked by cupules laboriously pounded into its surface, now covered with lichens, scat, and moss. No one really knows who made the holes or what they were for, although there are theories that they had something to do with fertility. Women may have eaten the rock dust, or perhaps the pounding itself was a trance technique. "As someone who spends a lot of time in trance," I said, "I have to say that there are simpler and easier techniques. I suspect that if you used this one, you wanted to do something with the product, whatever it was."

The boulder was perched above the stream, and Caerleon said most of the pitted rocks that have been found are in places where you can look down on something. They were often associated with fishing, she speculated. Below us the stream ran clear. No steelhead spawn in it today, but as late as the fifties, the locals say, you could fish them out with a pitchfork.

Gone, gone, gone.

The stream would have been larger before the logging—and more trees mean more water. The grasses in the meadow would have been different, not these European annuals that go dry and brown in the summer but deep-rooted bunchgrasses that stay green all year. The road wouldn't have been there, of course, but undoubtedly a trail would have followed the stream.

Where you find sedge, Caerleon said, it was probably planted by the Pomo for basketweaving. I told her about the talk I'd had with Victoria, one of the Pomo teachers at Living History Day at Fort Ross. I had told her that I was interested in basketry plants. She'd showed me sedge in their display, and told me that it's difficult to find, as so many of the traditional gathering places have been built over or bulldozed under. But for baskets, you need to find the sedge that grows on sandy loam, where the roots can spread out straight. On our rocky land, they tend to twist themselves into knots and gnarls.

Nevertheless, I've been transplanting a little sedge into my stream, for the spirit energy if nothing else.

After the walk, I go over to Jim and Dave's to dig some chestnut seedlings. I take them some bee balm and some lamium—White Nancy—which have been spreading nicely in my garden. With all the failures and frustrations gardening involves, there's still the something-for-nothing satisfaction of plants that propagate themselves. Lamium is putting down roots from every node, so I've now transplanted it all around the central circle and hopefully will soon realize my vision of a perfect full moon of silvery plants in the center of the garden, punctuated by white foxgloves, white lilies, white daffodil, anemones, and geraniums.

I never leave Jim and Dave's empty-handed. They've lived on the land for twenty years, and their house is surrounded by things that happily grow and spread and reseed. And right now, in mid-December, we are at Prime Time for Propagating. Cuttings rooted now will have months of rain to establish their roots.

In just the last few days, I've given them lamb's ears and red and purply-red penstemon, but they've given me purple penstemon, Sonoma sage, coreopsis, red and yellow obedient plant, horehound, many different lavenders, a beautiful succulent, coyote mint, mother of thyme, and a dozen carnation poppy seedlings, as well as the chestnuts. They've loaned me a Havahart trap to attempt to catch what I think may be a wood rat gnawing its way between our alcove ceiling and the roof, and taught me how to use a hoedad—the tree-planting tool par excellence. A hoedad has a long, narrow blade attached to a stick—a cross between a pick and a narrow shovel. You swing it into the ground and then step on it to drive it in deeper with your body weight, wiggling it in as you go. It makes a deep, narrow cut, perfect for inserting a slim seedling with long roots.

Dave assured me it would be easier to plant the chestnuts with the hoedad, and he was right. I quickly planted eight or nine seedlings, tying ribbons on so that I could find them again later to fit them with deer protection in the form of a small wire fence or plastic protective tube. I planted three or four on our water tank knoll, two or three on the far side of the garden, the others on the bank below the road. I had time to return the hoedad and climb Firehouse Hill to watch the sun set over the ocean. The work went quickly because, as I admitted to Jim, I've grown jaded: I just planted them; I didn't pray over them for half an hour this time or invoke the chestnut deva (or guiding spirit). But I do believe, after last year's failure with the dozen or so seedlings I planted, I have more of an instinct for where they might thrive.

Mother Earth

The earth is our mother, we sing. Mother earth, mother nature—she is the literal womb of life, providing all that we need. Her living soil feeds us; her rocks make our bones; her minerals are in our life's blood. The very heart of Goddess spirituality and of other indigenous traditions is the recognition that the earth is sacred.

This understanding of the earth is very old and very widespread. The earliest works of human art are ancient figurines of full-bodied, big-breasted, and big-bellied women that embody the sacred quality of life-giving earth/flesh. The painted caves of southern France and northern Spain were the living, sacred wombs of the mother, generating the animal hordes painted so vividly on the walls. In Greece, Gaia was the eldest of Gods. The life-giving Regeneratrix underlies the later Goddesses and compassionate mother figures, from Isis in Egypt to Kwan Yin in China to the Virgin Mary. The earth mother is mountain: the paps or breasts of Anu in Ireland, the "sleeping lady" who is seen in the volcano of Popocatepetl in Mexico and Mount Tamalpais in the San Francisco Bay Area. She is the mountain in the Himalayas that we call Everest but whose true name, Chomolungma, means Goddess Mother of the World.

The understanding of the earth as a living, sacred being is also very long-lasting. Archaeologist Marija Gimbutas, who did some of the major work on the early Goddess cultures of ancient Europe (as noted in Chapter Two), described in an interview how the peasants of her Lithuanian childhood in the 1930s used to kiss the ground every morning. And the reverence continues among indigenous peoples and those who work with the earth. As an example, at a recent meeting in Mexico City of Via Campesina, the worldwide small farmers' organization, began with an *ofrenda*, an altar/offering made on the ground of seeds, flower petals, and fruits of the earth to honor the earth mother, and a ceremony to exchange seeds.

And yet today, we also live in a global culture that profoundly dishonors the earth. Words associated with the earth are used as insults: "low," "dirty," "soiled." The ecofeminist movement of the eighties was founded on the insight that the way our culture treats the earth and the way it treats women are linked. Both are identified with the flesh, the body, the bloody and messy processes of bringing life into the world and its inevitable end in death, decay, and rot. When that cycle is devalued, when what is sacred is abstract, removed from earth, transcending life and death without being marked by the cycles of life, the earth and women are both denigrated and both become victims of exploitation, assault, and rape. So, too, those who live and work close to the earth are devalued and exploited. The farmer, the peasant, the manual worker, all are "low-class" workers. Manual labor is "beneath" the dignity of "high-class" folks, who work with their minds—or better yet, don't work at all, but live off of that ultimate abstracted value—profit—that is accumulated by the labor of others.

Our spiritual and ideological rupture from the earth is reflected in every aspect of our culture, but perhaps most deeply and ominously in the way we grow our food. Industrialized agriculture, the Green Revolution, biotechnology that produces genetically modified food plants, these are all based on a mechanistic understanding of soil and growth. Soil is just a medium to support plants, in somewhat the same way as the woman's womb was believed by Aristotle and the medieval Church to be only a vessel in which the male seed, the true life germ, was nourished. Plants can be fed a few key nutrients and protected from the competition of weeds and the predation of bugs by a blanket killing of anything that is not the desired crop. The true purpose of corporate, industrial agriculture is not to grow food, but to grow profits.

This form of agriculture has become one of the most destructive human activities on the planet. In the United States, we lose six tons of topsoil for every ton of food produced. We use thirty calories of fossil fuel for every calorie of food produced.[1] And we pour hundreds of millions of tons of toxic chemicals into our air, soil, and water.

Corporate globalization is industrial agriculture on a worldwide scale. Its vision is a world where no one will eat the food that they produce or food that is locally grown. Instead, food will be just another commodity circulated on world markets, generating more profit each time it changes hands. In the last two decades, we've lost a third of our family farms to policies that support globalized agriculture.

From the point of view of profit-making, an apple grown on a corporate farm in New Zealand, with the use of pesticides and heavy machinery, picked green (and later gassed in order to ripen), waxed, irradiated, and shipped in cold storage to be sold in a giant supermarket chain in California, is a great profit generator. It produces revenue for the farm corporation, the makers of pesticides and farm equipment, the food irradiators and packagers, the truckers and shippers and haulers, and the supermarket chain. Of course, those who actually till the ground, prune the trees, and pick the fruit receive only a pittance. According to the Oregon Department of Agriculture, out of every dollar we spend on food in the U.S., farmers receive only twenty cents. In 1950, they received forty-one cents. As late as 1980, they received thirty-one cents. No wonder small farmers are going bankrupt! And if farmers receive so little, farmworkers and migrant laborers receive far, far less.[2]

Perhaps a mother buys that New Zealand apple as a healthier alternative to potato chips for her child's lunchbox. Unfortunately, the poor apple is devitalized from its various treatments and its intercontinental journey, and laden with the residues of pesticides, herbicides, wax, and radiation.

Sonoma County, where I live, used to have a thriving crop of Gravenstein apples, a variety well suited to local conditions. Imagine the difference in the apple if that same mother, instead of selecting a New Zealand apple at the supermarket, bought fruit grown on a neighboring farm, organically, picked at its point of ripeness, and sold at a nearby farmer's market or neighborhood store. That apple would generate a different kind of benefit—increased health, taste, vitality, joy. It would create less profit for big corporations and chemical companies, but more real abundance for small family farms and local stores. And it would conserve some of the diversity of crops that makes for real food security.

Earth-honoring agriculture would generate abundance, but its primary intention would be not to grow profits, but rather to grow soil—living, healthy, complex soil—as a fertile matrix for living, vital, health-sustaining food. To grow soil, we need to appreciate and understand that soil is a living matrix of incredible complexity, the product of immense cycles and great regenerative processes.

Soil scientist Elaine Ingham lists what we might find in live, healthy soil:

> What do we mean, organism-wise, when we talk about soil? Agricultural soil should have 600 million bacteria in a teaspoon. There should be approximately three miles of fungal hyphae in a teaspoon of soil. There should be 10,000 protozoa and 20 to 30 beneficial nematodes in a teaspoon of soil. No root-feeding nematodes. If there are root-feeding nematodes, that's an indicator of a sick soil.
>
> There should be roughly 200,000 microarthropods in a square meter of soil to a 10-inch depth. All these organisms should be there in a healthy soil. If those conditions are present in an agricultural soil, there will be adequate disease suppression so that it is not necessary to apply fungicides, bactericides, or nematicides. There should be 40 to 80% of the root system of the plants colonized by mycorrhizal fungi, which will protect those roots against disease.[3]

EARTH OBSERVATION

Do this exercise not with your eyes, but with your nose. Ground, center, and come into your senses. Now, put your face in the earth and smell. Breathe in the fragrance of the living being that she is. The earth breathes, taking in air to fill her pores. And she breathes out, exhaling bits of herself that communicate her state of health or disease, her level of vitality.

Take a walk through the woods or through a park. Periodically, bend down and smell the earth. How does the smell change under trees? In the grass? What does the earth smell like where it is hard and dry and compacted? Soft and fertile? Can you notice different soil types, different histories? How does your own garden smell?

Get in the habit of sniffing the earth wherever you are. Get low, close to the ground. Learn the smell of healthy soil. (Just be careful on lawns drenched with chemicals and in areas that may contain toxins.)

The Cycle of Rock to Life

Life is rock rearranging itself.

—*Elisabet Sahtouris*

Let's follow a molecule of calcium through perhaps the longest of the elemental cycles. Here is a big, prickly leaf of the herb comfrey in my garden, which I've pulled up and laid down for mulch. Red worms, bacteria, and fungi have begun the process of decay. The rain comes and leaches away a tiny bit of calcium. That calcium could have many fates. First, it could be taken up again by roots, incorporated into the body of the lettuce I grow. From there it could end up in my dinner (later strengthening my bones), or it could be drawn up into that lettuce and my young neighbor Angie could eat a leaf, passing the calcium through her breast milk to her daughter, to build Ruby's baby teeth. Eventually Angie and the baby and I will all die, and our bodies will go back to the earth, and that calcium will return to the ground that it came from.

But let's say that it isn't taken up into a living body, but dissolves into the soil and passes down through the soil into the groundwater, seeping under the hills to emerge from a stream, and flowing with the water all the way out to sea.

Suppose it is taken up by a tiny, one-celled alga that is eaten by a shrimp that is eaten by a fish that is eaten by a bigger fish. And if it is eaten, at last, by a salmon or a steelhead, it might circle right to its source, via the salmon's migration upstream to spawn and die in the headwaters of the stream of her birth. And a bear might come and eat the salmon, and return its elements, including that calcium, back to the soil of the forest, closing the circle.

But, alas, our Sonoma streams are degraded, our salmon are gone, no bears roam our woods, and the forest is hurting for calcium and phosphorus and the smell of rotting fish.

But let's imagine that our lucky calcium makes it to the ocean and escapes being eaten by that alga, instead becoming part of the body of a tiny radiolarian, a one-celled organism that constructs an intricate, fairylike shell that looks like a spherical snowflake. And that radiolarian escapes the jaws of whales and the thousand other hungry mouths in its vicinity, to live out its natural life and die, letting its shell drift back to the ocean floor to be buried under a constant slow rain of detritus.

Our calcium lies there for a long, long time. Shells and skeletons fall from above, and the great weight of the ocean presses down, down. Slowly, slowly, our calcium is crushed down into the rock below, merges with rock, *becomes* rock, a formation of limestone weighing down the edge of the earth's crust.

And after millions of years, that edge begins to sink, to push down at the rim of the ocean plate, to churn up rock and magma from below, like a plow churning up the ground, pushing up a swell of magma, squeezing up mountain ranges, grinding and slipping along the fault line. Until at last the fault gives way, the edge of the plate lurches, cities fall, and our calcium is pushed upward, lifted high above the churning currents of molten rock to become part of a limestone plateau.

The rain will wash over her. The wind will caress and buffet her face. Lichen and ferns will begin to grow, and all together they will weather down that new rock to soil. And then our particle of calcium can begin, again, a journey through life-forms, leaf and bone to rock, from rock to life. And the circle is complete.

Many scientists today believe that it was the presence of life, the weight of those billions of radiolarians and other tiny skeletons pressing down on the tectonic plates of the earth, that set those plates in motion and sent the continents off on their long, slow perambulation of the globe. The cycle of life to rock to life is probably one of the longest-lasting of the earth's great regenerative cycles. The bones in the hands that are typing this sentence are strengthened by the bodies of ancestors hundreds of millions of years old, and for that I am grateful.

FERTILITY AND DECAY

In a world in which the life of the soil is everywhere under assault, building soil fertility can be a profound act of worship. To enhance the life of the soil, we first need to understand it. As Witches, we will probably learn more from a guided journey than a long, technical explanation.

Find a safe space—ideally, outside under a tree, lying on the earth. But you can also do this inside, if necessary. Ground, come into your senses, and create a sacred space. One person can read the following meditation for the group, or you can tape it ahead of time. Even better, the journey leader can read it through enough to become familiar with it, then make it her own, using her own words and images.

Lie down, breathe deep, and relax. Take a few moments to relax completely, from your head down to your toes. Feel the weight of your body on the earth, and feel the earth as a living body, embracing you, holding you close. Feel the elements in your bones and flesh that come from the earth.

And when you are ready, take a deep breath, and for a moment, imagine that you are a leaf, hanging tight to a twig on a high branch, waving in the wind. Feel the wind and sun on your face; feel how when the sun hits you your very cells sing with the energy of light, and the chiming chord they make creates a sweetness that permeates your blood and feeds your tall and reaching body. And just for a moment, let yourself hang in the breeze, feeling what it's like to feed from light, effortlessly. And if there is someplace in your human life that you need that feeding, that nurturing, take a breath and take it in.

And now time passes. Imagine the first cold winds of winter beginning to blow. And you feel them touch your face, and a freeze comes into your veins, and you glow scarlet in the light. And you take a deep breath, maybe a sigh, for you know that the summer has been good, but now it is past, and the time of singing sweetness is done. And if there is something in your life that is complete, some phase that is ending, something sweet that you now need to let go of, take a deep breath and draw in that power of the leaf to let go. And you let go and fall, letting the wind take you, and you swirl and dance and spiral in the wind's eddies, always falling, falling down and down and down.

Until at last you come to rest on the earth, lying on your sister and brother leaves, piling body upon body. But that earth you rest on is no solid barrier. It is porous, like a sponge, a labyrinth of cells and spaces, alive with a billion hungry beings. And you take a deep breath and give yourself back to the earth, and she reaches up to embrace you, and a billion hungry mouths open wide to take you in.

The ants and the beetles come up from below and begin to break you down. They eat away the soft parts and take apart the veins. Thin threads of fungus hurl themselves across your face, beginning the process of dissolution. You are drawn down, down, into the spaces and the caverns far below. Parts of you are ingested, becoming ant or beetle for a little while, then released again as frass that tumbles down into the earth. Parts of you are held in the fungal threads and slowly dissolved. Parts of you are licked by filmy mouths of soil bacteria, and slowly, slowly you are brought back into your original elements, and slowly, slowly you are brought down into the earth.

You descend, through great caverns and chasms, past great suspended archways of crystalline rocks, over sharp-edged, gleaming silica boulders and round, smooth spheres of clay. And the great caverns within the earth are slick with water, and in the tiny pools that form in her crevices a billion creatures swim. This is a whole, rich, three-dimensional world, and you are just a tiny grain of life, one speck of luminous phosphorus down here below.

Now out of the dark spaces beyond comes a great, smooth, writhing being, slick and wet, opening its great mouth to take in chunks of the very rocks themselves, opening new tunnels and pathways. The great worm meets another, and they slide along each other's bodies, sharing the liquid lubricant of their sweat, drunk on each other's odor, coupling at last in a doubled mutuality of instinctual pleasure, each one both male and female, each fitting to the other in a double lock. You are taken in, you become part of that shuddering mating, and then you pass out again, in a casting rich with your brothers and sisters and fragrant with the promise of life.

And now something reaches for you. Thin, thin threads, long arms of the mycorrhizal fungi that stretch between the root hairs of the great trees. Sticky as spiderthread, they snake through the caverns of the soil, holding the archways and the boulders in place, wrapping them in a living binding.

And all around you now, the caverns are penetrated by the most delicate, pearly, iridescent tubes of the root hairs of the plants and great trees above. And each gives off its own fragrance, its sweetness, its unique taste. And colonies of soil bacteria, those dancing circles, surround each one—each root attracting its own clientele like a street full of ethnic restaurants, each patronized by its own fans. But the trees—they throw out these tendrils of fungi that have got you lassoed. And you feel how they link root to root, how they can pass you from one tree to another down this network, how the trees feed their young, how trees that grow in the sun will feed trees who grow in the shade. Passing nutrients and energies through this phosphorescent web that you are now a part of, like the nervous system that feeds your brain. And you take a long, deep breath, and you listen to the long, slow thoughts of the trees.

And you know how the trees speak to each other, deep below the earth, and how the forest is linked through this web. This is the web that supports the forest mind as your nervous system and brain support your mind. And feel how far this web once stretched, when the forests covered the land. Feel what still exists, and what is broken; what still speaks, and what has been silenced. And know that you are held in this web.

And then, taking a deep breath, feel yourself sucked into the root hairs, and up into the root, rising and rising on a current of sweet sap, rising up now and caught in a great upward tide. Higher and higher, through the channels within huge roots and living skin beneath the bark, and out into the branches, the twigs. Where you become part of a green bud that opens with the warmth of the spring sun, unfurling itself like a wing to catch the sunlight and sing it into sweetness.

And there you wave, a brave banner, a signal flag of life, through the warm days and the long nights, catching sunlight, singing sweet food out of air and water to feed the twigs, the bough, the trunk, the root. Until once again the cold of winter comes, and again you will let go and fall to earth, to be taken in, to be brought back down to your original elements, to sink, to be eaten, to feed the roots that feed the trunk and the boughs and the twigs where leaves cling for a time, singing sweet food from sunlight, until the time comes to fall to earth, to feed the roots, to grow the trunk to hold the boughs to carry the leaves to sing sweet sunlight into food until they fall again, to feed the roots, to grow the leaves, to fall, to sing, to feed, to grow, to fall, to sing, to feed . . .

And you breathe deep again. And you begin to remember that you have a human body, a human mind. And you thank the tree, the leaves and the roots, the fungi and the worms, the billion hungry mouths of the earth, for this journey, and for what you have learned about the cycle.

And slowly, slowly you begin to breathe yourself back into your human body, feeling your feet and legs and torso and arms and hands and head take shape again. Slowly begin to move and stretch, to feel the edges of your human body. Slowly begin to sit up, to open your eyes. Say your own name out loud. Clap your hands three times.

And that's the end of the story.

Decay Is Food

The story above teaches us one of the great lessons of both earth-based science and spirituality—that there is no such thing as waste. Waste is food. All fertility arises from decay. There is no life without death, and death feeds new life. Life and death, decay and regeneration, are part of the same cycle. We cannot have life without death, fertility without rot. But death need not be feared or viewed with horror. It is part of the cycle—a transformation, not an end.

According to Marija Gimbutas, the most ancient Goddess of Europe was, beyond all, the Regeneratrix, the one who brings fertility out of decay. We serve her whenever we take responsibility for our wastes, returning them to the cycle of fertility.

One of the simplest ways to do that is by making compost from our organic food wastes. My first coven was called the Compost coven, and a descendent of that group still exists. I have always considered compost to be sacred.

Lazy Compost

There are volumes written about various ways to make compost, and the sale of a variety of different designs for composters is a small industry. Generally speaking, there are two basic approaches to making compost, and these correspond very closely to the two basic approaches to magic. The first is the Ceremonial Magician/Alchemist School (of composting or of ritual), which involves lots of complex processes that must be done at exactly the right time, and many careful measurements. I am not going to discuss that school here, because it is not my form of practice. If you are temperamentally drawn to such things, however, they can certainly have enormous value. I encourage you to look into biodynamics, for which many good resources exist.[4]

The second school is the Kitchen Witch/Permaculturalist School, which uses whatever you have lying around. It's the method for lazy or busy people, which is why it appeals to me. (I'm not saying which I am!) I've always believed that if there are two ways of doing something, and one involves much less work than the other, that's the way to do it.

There are a number of different approaches to the Lazy School of compost-making, but all of them hold in common the essential insight that when it comes right down to it, things rot. In making compost, we're not only working with nature, we're hastening what nature would do anyway, whether we interfered or not.

So . . . the essence of the lazy method is to pile up your wastes and let them rot. There are, however, a few simple tricks that will help material rot gracefully, odorlessly, and reasonably quickly.

We can think of a compost pile as made up of two basic sorts of material, which for simplicity's sake we'll call "green" and "dry." Green stuff is high in nitrogen. It includes food scraps, fresh grass clippings, fresh-pulled weeds, etc. (It also includes manure, although that tends to be brown.) Dry materials are high in carbon: dry leaves, straw, newspaper—your basic brown and crunchy stuff.

A successful compost pile has a rough ratio of carbon to nitrogen that is thirty to one. That's right—thirty! The microorganisms in the soil that break down plant materials need about thirty times as much carbon as nitrogen to thrive.

Because green stuff also contains carbon, which is the basic structural element of life, in practice what you need is about half green and half dry. Many of the problems people have with their compost pile—smell, flies, not breaking down, etc.—can be solved by topping the pile with more dry stuff.

The other big secret is to build the pile *big* enough and *damp* enough. Rotting material generates heat, and a hot compost pile will kill weed seeds and undesirable microorganisms. To get hot enough to accomplish this, a compost pile needs to be at least three feet in diameter and about that high, and it needs to be damp—not too wet or too dry, but about the consistency of a damp sponge.

So the very simplest way to make compost is to collect enough wet and dry stuff to layer it up three feet high; then keep it damp and let it rot. If you build your compost pile on top of the bed you want to plant, you will even save yourself the trouble of later transferring it.

A mass of compost will rot down amazingly quickly. I once nobly took an entire pickup-load of rotting food wastes back home with me from Witch camp. With the help of the other teachers, who had accompanied me back to my land for what they thought was a retreat, we made a pile of food scraps layered with straw in a bin two feet wide, four feet long, and three feet high. Within a few days, the pile was a third that high. Within a week, it was half its original size. In a month, it had virtually disappeared.

Some people turn their compost piles, to aerate them and help them break down faster. You can do that, if you like or if you need the compost quickly. Essentially, you are trading work for time—more effort for faster results. Alternatively, you can leave it alone, to rot peacefully undisturbed, trading time for work.

If you live in the city, it's worth investing in one of those black plastic composters that will keep rats out of your pile. If you can't afford one, get three or four old tires. Put down a base of wire of small enough gauge to keep rats out, and affix it to the first tire. Stack the others, and pile the compost inside, with a board on top and a brick or large stone to hold it. If you make two compost piles, one can be "cooking" while you keep adding to the second.

COMPOST BLESSING

We offer gratitude to the great cycles of birth, growth, death, decay, and regeneration. We are grateful to all the beings who have made the great transformation, leaving the remains of their bodies here. We are grateful to all the hungry mouths that consume the dead. Blessings on the termite, the beetle, the ant, the spider, the worm. Blessings on the fungi and the bacteria, those that need the air and those that avoid it. Blessings on all the life in this pile that will transform decay to fertility, death to life. May I always remember that the cycle of life is a miracle. May I continue to feel a sense of wonder and joy in the presence of death and life. May I remember that waste is food, and may my eyes be open to opportunities to close the circle and create abundance and life.

COMPOST PILE SPELL

I find my compost pile to be a good ally in dealing with problems that feel stuck to me. Here's a simple spell (best done while the moon is waning).

Ground, center, create a circle, honor the elements, and bless your compost pile. In your circle, have a large piece of fruit or vegetable ready to be discarded, and materials to write or draw with, either on the fruit or on paper. Take a moment and think about your problem. You can write it out on the paper and stuff the paper into your fruit, if the fruit is large enough. Or you can carve or draw directly on your object. Hold your fruit and breathe into it, imagining that you are letting the stuck energy of your problem flow into its flesh.

Say, "I give this problem, this energy, in the body of this fruit [vegetable, apple, carrot, etc.] to the great cycle of birth, death, decay, and regeneration. May it decay in its present form, and be brought back to its essential elements. May I see those elements clearly. May it fertilize some new seed, some new growth. With the offering of this fruit, I give thanks to the cycle, thanks to the processes of life and growth. I give thanks for this transformation. Blessed be."

Now bury your fruit in the compost pile.

Thank all the powers you've invoked, and open the circle.

Sheet-Mulch—Even Lazier!

Permaculturalists employ an even easier method of creating fertility, called sheet-mulch. Essentially, we turn the whole garden into a compost pile, spread out horizontally. With sheet-mulch, we are also trading time for work. A sheet-mulched garden may take longer to establish itself than a carefully dug raised bed, but it will save you all that backbreaking digging. If you don't want to have to wait, you can dig a small area for a prize vegetable bed and sheet-mulch everything else.

One advantage of sheet-mulching is that you don't disturb the complex structure of the soil or interfere with all those billions of happily munching bacteria. Mulch creates habitat for worms, who will aerate the soil for you. But if your soil is seriously compacted and dead to begin with, digging once to aerate and break it up may be helpful.

The hardest part of sheet-mulching is collecting the materials—lots of them! Cardboard, newspaper, old rugs, and even old clothes can be used for sheet-mulching. You'll also need lots of both green and dry materials.

Push down the weeds in your garden bed. (You don't even need to cut them; just press them down well.) If you want to jumpstart the decay process, add some high-nitrogen material. This step can be as simple as peeing on the pressed-down greens! (Human urine is about eighteen percent nitrogen.) Then cover the greens with cardboard or thick layers of newspaper, about a section thick. Don't use high-gloss, colored parts of the paper. Today's papers are printed primarily with soy-based inks and should be safe to use. If your site slopes, work from the top down so that lower pieces of cardboard lie atop higher pieces, to catch water—a reverse shingle effect.

Then add a layer of green stuff, and cover again with a layer of dry stuff (the thicker the better). It's really as simple as that. Water it all down to dampen it and start the decay. If you want to plant right away, just punch a hole in the cardboard and plant down through the hole.

As the sheet-mulch settles in, it will attract worms and soil bacteria. Keep mulching. Tuck your food scraps under the mulch and they will compost quickly.

Chicken "Tractor"—The Laziest Method!

If you live in an area where you can keep chickens, you can dispense with compost and let the fowl do your gardening. Simply create an enclosure for the chickens that covers the area you intend to plant. Feed them your kitchen scraps and let the chickens dig, weed, and manure the area while eating the pesky bugs. At the same time, those chickens will provide you with eggs.

When the ground is prepared, move the chickens and plant the bed.

Worms

If you are a single person or a small family composting in a small, urban back yard, you may have difficulty keeping chickens or gathering enough material for a true compost pile. For you, I recommend worms.

I love worms! Worms get a bad rap: "Lowly as a worm," "Nobody likes me, everybody hates me, I'm going to eat some worms," etc. In reality, worms are the least lonely of creatures. Each is both male and female, and whenever worms bump into each other in crowds, they mate in writhing balls of slithery worm orgies, indiscriminately fertilizing and being fertilized simultaneously. No constricted gender roles for them!

Along with furnishing you with the vicarious enjoyment of their erotic exploits, worms are the great creators of fertility. They tunnel into the soil, turning and aerating it. They eat soil particles and rotting food, passing them through their gut and turning them into worm castings, an extremely valuable form of fertilizer high in nitrogen, minerals, and trace elements. They add soil bacteria to the mix as well, inoculating the garden with many helpful bacteria.

It is the surface-dwelling red worms that eat food scraps and waste, along with manure. Some garden stores or bait shops sell worms, but you can often find them in your compost pile or around half-digested wastes on the ground. To thrive, worms need food, moisture, and a temperature that's neither too hot nor too cold—50s through 70s Fahrenheit being ideal.

You can create a bin for your worms by drilling some holes in a plastic bin or a plastic garbage pail for air and drainage, then putting in some earth, some shredded newspaper, and some food scraps. Add your worms, and let the whole thing settle in for a couple of weeks. It may go through a disgusting phase, where everything molds and rots (especially if you added a lot of food at first). But eventually it will settle down, the worms will start chomping their dinner, and you can start collecting lovely black worm castings.

You can add the castings directly to the ground around your plants, if you want, but I like to make "worm tea," dissolving a handful of castings in a bucket of water and then using the water to fertilize the plants. At times, I can almost see the plants perk up and lick their lips.

My own preferred way to keep worms is in a "tower"—a masonry chimney pipe just a bit wider than the five-gallon bucket it contains. The pipe sits in the center of my garden bed. I have a worm colony in the bucket, with drainage holes in the bottom. From time to time I wet down the colony with the hose: the water drains through and inoculates the bed with worm tea, soil bacteria, and eggs. Sometimes I put another small bucket below the colony's bucket, fill it from above with water that drains through the top bucket quickly enough that the worms don't drown, then pull up both buckets and use the tea from the bottom bucket to water plants.

The masonry chimneys are also attractive to slugs and snails, who congregate inside, where they can be easily picked off and, after a short prayer to Kali, dispatched.

In a cold climate, you can take your worms into the garage or basement for the winter. My friends Lisa and Juniper keep their worms in the kitchen. They have three plastic bins, stacked at a slight angle. Each contains worms, shredded paper, and food scraps. They keep the worms moist, and the worm juice runs down into the lower part of the bins, where they can siphon it off with a turkey baster and use it to water their houseplants.

In Sebastopol, California, the RITES project (Returning Intention Toward Ecological Sanity) collects food scraps from the local restaurants and raises worms on a large scale. They are now selling their worm tea as fertilizer, to fund some of their programs.

Worms truly represent abundance. Last spring, I was redoing my garden beds and went to shovel out our compost bin. Worms had gotten into it at some point, and it had turned into a mass of worms and worm castings three feet in diameter and three feet high. I subdivided them into new worm colonies, started new buckets full to give away, and still had enough left to cover all my major garden beds with a couple of inches of worm castings. Then I went out to the garden store to pick up a few plants. I noticed that the store was selling a pint container of rather depressed, pathetic-looking worms for $13.95. I figured that at that rate I had just put about $10,000's worth of worms onto my garden, and I started to rethink my whole career!

And the garden has indeed exploded with fertility, with tomatoes and corn and flowers crowding all over each other. Whenever something starts to look a bit peaked, I just give it a dose of worm tea.

And although eating worms sounds unappealing, they are actually a good source of protein. Although I must admit I haven't eaten any yet, I derive a slight sense of security from knowing that I *could*, if the worst happened.

One caution with worms: don't dump your worms into a pristine wilderness environment. There is some evidence that exotic worms in northern forests, whose own worms went extinct during the Ice Age, can disturb the ecological balance and *destroy* fertility instead of *creating* it.

Fungi

Fungi also are not generally well looked upon. Yet more and more we are coming to understand the critical role they play in fertility and regeneration.

Some forms of fungi break down the tough chemical bonds in wood that keep it strong. Others are important in the general decay of plant material. Still others are symbiotic with living plants.

The mycorrhizal fungi that we met in our trance journey are a vitally important part of forest ecologies. Threadlike and spreading, they insert themselves into the root hairs of trees and almost all other plants (except grasses). They then extend the reach of the roots, drawing in more water and nutrients than the plant can reach alone, in exchange for sugars extruded by the roots. The network of sticky threads helps hold the soil in place and allows plants to communicate and share nutrients. Through the fungal network, trees can nurture their young. Trees in the sun will feed trees in the shade—even those of a different species. A clearcut forest, where the mycorrhizal fungi have died, will not easily regenerate.

Other sorts of fungi, including mushrooms, are also symbiotic with trees. The part of the mushroom we eat, the fruiting body, is only a small extrusion of the whole organism, which exists predominantly underground and often interpenetrates the roots of living trees. The mycelium, the mass of threadlike tissue that constitutes the main body organism, extrudes its fruiting bodies when conditions of temperature and moisture and disturbance are right. Collecting wild mushrooms is a specialized hobby, and the scope of it is beyond this book. There is no simple rule of thumb for telling poisonous from safe varieties, and a mistake with mushrooms is one you don't want to make. But with care, wild mushrooms can be harvested. Eating the mushrooms I find in the woods is, for me, truly eating the flesh of the land. Each winter I pick and dry the varieties I am sure of: king and queen boletes, chanterelles, and matsutakes.

Mushrooms are also great healers. Chinese medicine recognizes several varieties as medicinals—reishi, shitakes, and many others. In a good mushroom year, when I eat them through the winter, I get fewer colds and flu and have abundant energy. Turkey tail mushrooms, which grow on old logs, are a natural antiviral and antibiotic when steeped or chewed.

Paul Stametz, author, mycologist, and founder of the company Fungi Perfecti, is truly the wizard of mushrooms. He has shown that mushrooms can also be healers of the earth. Oyster mushrooms will break down diesel fuel and other toxins. Stopharia can cleanse fecal coliforms from water. Shitakes and other varieties can be grown on cut wood as part of a program of truly sustainable forestry.[5]

The humble fungus certainly deserves more honor and respect from those of us who honor the earth. The fungus is a bit like the beggar in fairy tales, who appears lowly and dirty but offers great gifts and wisdom to those who treat her with love and generosity.

Sacred Seed

At the Via Campesina ritual mentioned above, an indigenous healer spoke of the sacredness of the seed. "The seed is sacred, because the seed is the beginning of all life," he said. "Everything comes from the seed."

Seed is indeed a sacred trust. The seeds for all of our traditional food plants are a precious gift of the ancestors, who saved them, selected the best each year and put them by for the next year, over centuries and millennia of time breeding thousands of different varieties of food plants adapted to different conditions, climates, and soils, and offering different advantages.

Seeds also represent abundance. On my table sits a bowl of fava beans, saved from plants I grew. Each bean can potentially grow a new plant, yielding many pods, dozens more beans. Each seed contains the instructions for its own replication and multiplication.

Farmers and gardeners have always saved seeds, traded seeds, and gifted each other with seeds. Seed-saving and seed-sharing are part of a network of relationships that hold communities together just as the sticky threads of mycorrhizal fungi hold the soil.

Seeds are also libraries of genetic information. Each seed holds the whole history of evolution, the record of hundreds of thousands of choices and accidents. The DNA in a seed may express itself in traits of the plant, but each seed also holds unexpressed and dormant potentials that may sometime in the future give rise to new varieties. Each seed is a concentrated communication about how best to grow and live and die in a specific place.

SEED MEDITATION

Hold a seed in your hand. Close your eyes and breathe deep. Feel the life, the coiled potential that you hold. As you breathe, relax and let your mind become a clear pool. Ask the seed to speak to you, to show you its history and its wisdom. You may see the faces of the ancestors, one after another, reaching back through time—each woman or man who guarded this chain of life, growing the plants, selecting and saving the seeds. You may catch glimpses of other landscapes, other places.

Ask the seed what it needs and wants in order to thrive.

Thank the seed, the ancestors, the land, and the elements, and open your eyes.

SEED SPELL

If you've let your stuck projects and problems decay in your compost pile, you may be ready to start something new. Planting seeds is also a good way to plant new ideas or new projects. Do this in the garden, after preparing a bed.

Ground, center, create a circle, honor the elements, and bless your seeds (see below). Take a moment and think about your upcoming project or the new beginning you wish for. Hold a seed and breathe into it, imagining that you are filling it with the image of your new project and with all your enthusiasm and passion.

Say, "I ask help from the great cycles of renewal for this project. As this seed holds a world of potential, so too does the germ of my new project. May I see and realize its full potential. As this seed grows, puts down roots, and sends out shoots, may my project likewise grow. May it find the nutrients it needs to flourish; may it be well-watered; may it thrive. May it create true abundance and further the diversity and joy of life. As I plant and tend this seed, I give thanks to the cycles of life and to the great mysteries of birth, growth, death, and regeneration. Blessed be."

Thank all the powers you've invoked, plant the seed, and open the circle. Don't forget to water and tend your seed as you tend your new beginning.

SEED BLESSING

We give gratitude to the elements of life embodied in this seed, to the wisdom accumulated over billions of years. We thank the ancestors who made the choices that gave us this seed, and who tended the chain of life to preserve it. We thank the air, the sunlight, the water, and the earth that sustain the life of this seed. Within this seed are precious and unique instructions for growth and life. May we always treasure the wisdom of the seed, and may we have the help we need to continue to nurture this life for the future. Blessed be the seed.

Seeds in Jeopardy

Unfortunately, the precious heritage of seeds and diversity is threatened by today's agricultural and economic systems. Traditional seeds are "open pollinated." Open-pollinated seeds breed true—that is, their offspring are true to their parents, with only minor variations. Such seeds can be saved from year to year, crop to crop. Farmers and gardeners can save their own seeds and don't need to constantly buy new supplies from seed companies.

Hybrid seeds are the product of more genetically diverse parents. Generally, they do not breed true—their immediate offspring may be quite different from the parent plants. And many first-generation hybrids are infertile or less fertile. Over many generations, traits can become "fixed" in some hybrids, but this often takes time and professional expertise.

Over the past century, large seed companies took control of most of the world's seed production. They specialized in hybrid seeds, born of two genetically diverse parents that do not breed true. Many hybrid varieties have advantages over their open-pollinated ancestors, but they have one large disadvantage for farmers and gardeners: their offspring may be quite different from their parents, so new seed must be bought for the next crop. Of course, this is an advantage for seed companies, which reap more profit when farmers are forced to buy new seed each year.

Genetic engineering carries this process further. By contract, farmers who purchase GMO seed are not allowed to save it. Companies hire private security agencies and encourage neighbors to turn in violators in order to safeguard their royalties. As people turned away from gardening, and from saving and sharing seeds with their neighbors, becoming more dependent on buying seeds, much diversity has been lost. In *Seeds of Change*, Kenny Ausubel writes: "Of the cornucopia of reliable cultivated food plants available to our grandparents in 1900, today 97 percent are gone. Since the arrival of Europeans on this continent, 75 percent of native food plants have disappeared from the Americas."[6]

But in the past few decades, corporate globalization has mounted a campaign to control not just the seeds, but the underlying genetic information within them. International trade agreements and institutions such as the World Trade Organization have allowed the patenting of life-forms, and have enforced those regulations worldwide.

What does this mean? For centuries, farmers in Sinaloa, Mexico, have grown yellow beans. In recent years they have earned an income by exporting them to the U.S. Until a man named Larry Proctor took some beans, grew them out for two years, patented them, and began demanding six cents in royalties for every pound of beans sold. The importers simply stopped importing those beans, and the farmers of Sinaloa, descendents of the men and women who had developed the beans over millennia, lost 90 percent of their exports.[7]

In India, for thousands of years the neem tree has been a source of medicine, insecticide, and oil used for many products. Many small producers made a living from products made from the neem tree. Until a large corporation, W. R. Grace—together with the U.S. Department of Agriculture—patented it, forcing the small producers to close down. Thanks to the hard work of Vandana Shiva and the group Diverse Women for Diversity, the European Patent Office struck down this patent in 2000.[8]

This story, repeated over and over again in abundant variation, is now common throughout the world. It amounts to a form of biopiracy, a theft of the gifts of the ancestors and of the very genetic material that underlies life itself.

The theft of genetic material, in the context of a system that allows the patenting of life-forms, provides a basis that makes genetic engineering potentially profitable. The biotechnology of genetic engineering produces GMOs—genetically modified organisms.

Genetic modification is very different from the plant breeding that farmers have done for millennia. Traditional plant breeding means selecting parents that are similar enough that they can produce offspring, then selecting various of their offspring over many generations for desired traits. We can change the character of a plant—but slowly, over time, and within parameters that do not fall too far from the original characteristics. Any naturally occurring change involves the whole organism, and any flaws in the breeding will generally become apparent in the organism's failure to thrive. Selecting too assiduously for one trait may result in other, undesirable, traits being passed on. Roses bred for color, for example, may lose scent or vigor. Because whole organisms have to grow up to reproduce, any change is always being tested in the context of the survival of the organism and its relationship to the other organisms around it.

Genetic modification is something very different. It involves artificially inserting genes from one organism into another organism—one that may be entirely unrelated. Genes from a flounder have been put into a tomato, for example—presumably to increase its protein content. Bacterial genes have been inserted into corn to make its pollen toxic to insects. Corn plants have been modified to resist herbicides, so that they and everything around them can be blanket-sprayed.

"Genetically engineered crops represent a huge uncontrolled experiment whose outcome is inherently unpredictable," Dr. Barry Commoner stated in a report challenging the basic scientific assumptions upon which the technology is based. "The results could be catastrophic."[9]

We do not know the ultimate impact of the radical changes in organisms that genetic modification produces. And we do not know what the unintended consequences of these changes might be. When organisms are not changing as a whole in relationship to a whole, integrated community, but are being manipulated piecemeal, there is no safeguard of survival over time to ensure that changes will be beneficent. Earlier in this book, I described the near-debacle of *Klebsiella planticola*, the soil bacterium that unexpectedly prevented plants from growing. Luckily, that outcome was detected before its release. In a similar vein, pollen from insecticidal corn could kill butterflies as well as corn-borers, or trigger intense allergic reactions in humans.

Experimenting with such things in the laboratory might be of some value, but releasing these products into the environment is a huge threat to biodiversity and global food safety. Crops such as corn are wind pollinated, and there is no way to confine genetically modified pollen to one field. Genetically modified corn was introduced to Mexico in 1996, against many objections from farmers, environmentalists, and local indigenous communities. It has already contaminated corn crops in Oaxaca and Puebla, the region where corn originated and the center of corn's biodiversity. This contamination represents an irreparable loss. For when plant breeders seek to restore a lost trait or improve the vigor or health of a crop, they often cross back to varieties closer to the wild. If the biodiversity of the wild stock is compromised, we lose a huge amount of potential for developing new adaptations.

Moreover, once a crop is contaminated by patented pollen, farmers can be forced to pay the very corporations that have polluted their crops! Percy Schmeiser, a farmer from Saskatchewan, Canada, was sued by Monsanto after their genetically modified canola contaminated his crop. In fact, their pollen damaged his canola, but instead of Monsanto paying him, he was forced to pay them $19,000, and they sued him for back "royalties" for growing crops that include their genetic material![10] The Canadian Supreme Court ruled in Monsanto's favor, upholding their right to royalties on organisms that include their patented genes. However, Schmeiser did not have to pay damages, as they could not prove that he benefitted from having the genes in his crop.

The major crops that have been genetically modified—corn, canola, soybeans, and sunflowers—are all wind pollinated. They were chosen deliberately by corporate scientists precisely for their potential for contaminating other crops—for once a crop is tainted, the farmer can be forced to pay up and prevented from saving his or her own seeds.

Pharmaceutical crops are now being grown that include human genes. We do not yet know what eating these crops will do to human beings. We are only now, after decades of exposure, beginning to understand that pesticides and herbicides do cause cancer and other diseases. Human genes in foods could trigger allergic reactions or cause other unexpected health effects.

The proponents of GMOs often invoke the "free market" as a justification for their profiteering. They claim to be providing what the market wants. But, in fact, the market rejects GMOs wherever it is free to do so. Europe and many third world countries have resisted the introduction of GMOs. Mendocino County in California and many Vermont townships have also recently voted in bans. But once an area is contaminated, it loses one of its strongest arguments for banning them. "They're already here," the argument goes. "It's too late. Oh well."

The companies that produce GMOs have resisted any attempts in the U.S. to require labeling that would allow consumers to make a truly free choice about what they want to eat. A "free market" is one in which consumers have full information about the products they choose among. If we did have a choice, an informed public would be unlikely to make GMOs profitable.

The patenting of life and the creation of genetically modified foods are part of the privatization of nature, the attempt to turn nature into a commodity that can be bought, sold, and controlled. If we believe that nature is alive and sacred, beyond price, if we say that the heritage of knowledge and legacy of biodiversity created by the ancestors should be common to all people, we need to educate ourselves on these issues, and to oppose the patenting and genetic manipulation of life-forms.

Growing some of our own food, cherishing biodiversity, and saving and trading seeds are also ways we can honor the life forms and the legacy of the generations past. Supporting small-scale local and regional agriculture is a productive form of resistance to a profit-driven vision of a world in which no country eats what it produces, but gears everything to exportation.

We now know something about how to grow soil. Let's consider a few more ideas about plants.

Plant Communities

Abundance arises from complex webs of association and cooperation. In nature, no species grows entirely alone. Plants grow in community, in association with other plants. And in those communities, each fulfills certain roles.

Some provide shade and leaf litter that nurture others. Leguminous plants, those in the pea and bean family, fix nitrogen from the air into the soil, providing fertility for others. Dynamic accumulators, deep-rooted plants such as comfrey and burdock, bring minerals and trace elements up from deep in the earth, returning them to the areas near the surface where they can be used by others. Other plants, known as "insectaries," attract beneficial insects and pollinators. Among these are many common flowers, including asters, sunflowers, yarrow, thyme, and Queen Anne's lace. Still other plants concentrate different minerals in their leaves and stems and branches, returning them to the soil when they die and rot. And *every* plant's roots are a feeding station for fungi and bacteria, in ways we are only just beginning to understand. Birds and animals contribute too, bringing in manure, digging and pruning, and catching bugs.

In one of my fields stands a huge old Oregon oak tree. Beneath it are so many sprouting Douglas firs that I refer to the spot as "the conifer daycare center." Doug firs have been replanted in many areas of my land, and have come back naturally in others, but nowhere as thickly as under that tree. Clearly, something in its shadow or in the soil conditions it creates favors their growth.

"Guild" is a permaculture term for a self-sustaining mini-system of plants growing together in ways that support each other. Guilds are found in nature, or course. Across the road from my garden, for example, is a madrone/tanoak/gooseberry guild with firs coming back through the larger trees.

PLANT COMMUNITY OBSERVATION

In your home base, ground and come into your senses. Take a walk around and observe plant communities. What grows together with what? Do you have a sense of what roles each plant fulfills? Are some plants serving as nurses for others?

Are there patterns of association that repeat? Plants that seem to like each other? Dislike each other?

Is there a spot that's especially diverse? Any idea why? Are there certain conditions that favor certain plants?

You might want to make some notes for your journal on this one.

A Plant Guild

If we want the garden to be a self-sustaining system, we need to think about planting guilds, not just individual plants. There are many factors to consider. Let's take the case of my nectarine tree.

My nectarine tree grows on a windy hill with poor soil and produces small, wind-battered nectarines of incredible sweetness. It was planted by the former owner ten or twenty years ago, and grows outside the deer fence, away from the drip irrigation, in a field full of rough grasses, particularly Harding grass—a tough New Zealand native that was planted by mistake after the big Creighton Ridge Fire of 1978, when an aerial seed crew grabbed the wrong bag of seed. It is now the scourge of the fields around here.

What does the fruit tree need? In one sense, nothing—it's surviving if not thriving after years of complete neglect.

In another sense, what does *any* plant need? Sunlight, water, and various nutrients.

Sunlight is no problem for the nectarine. On a ridge with no obstructions to the south, it gets plenty of sun.

Water is abundant in winter and spring. In the summer, the nectarine's roots are evidently deep enough that the tree can survive our normal dry period. More water might produce more fruit, but that fruit might also be less distinctive, less concentrated in its taste. Trees are also a *source* of water. They concentrate water at their dripline and around their trunks, and their leaves and branches comb the fog for moisture. So the nectarine might actually provide some moisture for other plants.

My nectarine also needs nutrients. Nitrogen, phosphorus, and potassium are the three key nutrients, along with various minerals and trace elements. This tree has new growth in the winter, and it produces fruit, so it's not starving. Fruit trees in general don't need to be highly fertilized. Too much fertilizer can stimulate a lot of woody growth instead of fruit production. And its own leaves provide nutrients and mulch for other things I plant. Still, I might want to give the nectarine some companion plants to help feed it.

First, I might plant something from the legume family, to fix nitrogen—possibly lupines or a groundcover of clover. Peas or beans will simply get eaten by deer.

Then I could put in some dynamic accumulators: plants that concentrate minerals. Comfrey, borage, bracken, and other deep-rooted plants feed from a level below the roots of the fruit tree, thereby bringing lost nutrients back to the surface. In general, I like my fruit trees to have a comfrey "pet" to keep them company. When the comfrey leaves die and dry out, they also provide a good mulch.

The matted roots of grasses compete with fruit tree roots, so to keep back the grass I might plant a ring of bulbs that are drought-tolerant and deer-proof. Daffodils or irises would do nicely, and they would also help keep the soil loose and provide me with flowers in the spring—a small crop.

If I want wild bees and other pollinators around in the spring when the tree blossoms, I need to provide food for them all year round. In addition, I want to attract predatory insects that eat pests. So I might include some of the apiaceae, the family of small-flowered plants that used to be called umbels (as in umbrella, because of the form of their flower sprays). This includes Queen Anne's lace, fennel, parsley, and dill (although the deer might eat the latter). I'll also include some asteraceae, some plants from the daisy/sunflower/aster family—maybe perennial Mexican sunflowers or Michaelmas daisies.

I might also think about some night bloomers, to provide food for night-feeding insects, which in turn provide prey for bats and other birds. Soap lily is a native bulb with delicate, miniature lily sprays that open at dusk in the summertime. I could include some water catchers as well—plants with shiny leaves to condense water, or needles and filigree to comb fog. Other native plants provide habitat for insects and animals. And I might include some herbs—perhaps thyme, lavender, sage.

In putting together the guild that would support my nectarine, I have some other criteria. Everything I use must be drought-tolerant, requiring little or no supplementary water. It must also be deer-resistant, because I'm not going to fence this area. And it must be a light feeder, requiring little or no extra fertilizer, so as not to compete heavily with the tree or force me into rounds of fertilizing.

Everything in this guild should serve more than one function. I'm willing to include "looks pretty" as one function, but ideally every element in a system should serve at least three functions. So I might include a native ceanothus, which would fix nitrogen, bloom in spring for pollinators and beneficial insects, keep back the grass, provide habitat for natives, and look pretty. Thyme would feed insects, keep the grass back, and provide me with herbs to eat. Soap lilies would not only be natives, night bloomers, soil amenders, and grass repellers; their roots could be used to make a natural soap and shaped into natural brushes.

Many years ago, I did plant a guild very much like the idealized one described above. I began with a sheet-mulch around the tree to kill off the grass. I first cut the grass, then placed a layer of thick cardboard over it, and finally added a layer of soil that I dragged up from the bottom of a seasonal pond. I was worried that the usual sheet-mulch of manure and compost might be too much for this tree and for the plants I intended to use around it. Besides, I had the pond soil and didn't have the manure and mulch at the time, and one permaculture principle is to use local resources. I planted lupines, comfrey, thyme, lavender, irises, sage, dittany of Crete, and daffodils. All of them still survive, seven years later, as does my wind-blown nectarine, in spite of many changes and several summers of utter neglect. It has proved to be a self-sustaining guild.

Maybe this fall I'll take some time to fill in the circle with more of the various plants I've been considering, throw more sheet-mulch over the grass, and propagate the sage, a variety native to the Southwest, which thrives here and could provide a nice crop for smudge sticks. A garden can always use improvement.

Or maybe I'll just leave well enough alone.

PLANT ALLIES

Plants are always communicating with us, and we can learn to deepen and intensify that communication. When we make special friends with a plant or with a species, that plant becomes our ally. If we use allied herbs to heal ourselves, that alliance can intensify their curative powers. If we grow food or flowers, we can provide plants with spiritual as well as physical nutrition.

Plants like to be talked to, sung to, appreciated. Flowers are vain: they like to be admired. As Mabel McKay said, if we don't use the plants and talk to them, they'll die.

Here's one way I approach making an alliance with a plant:

In a safe space where your plant grows, ground and come into your senses. Create a magic circle, and bless the elements and the Goddess.

Sit with your plant for a bit. Take time to observe it with all your senses—to look, listen, smell, touch, maybe even taste (unless, of course, you're making an alliance with poison hemlock or something similar).

Sprinkle a few drops of waters of the world at the base of the plant as an offering.

Now sit, breathe deep, and close your eyes. Picture the plant in your mind's eye, and ask permission to enter it and make alliance with it.

If you sense that the answer is yes, imagine the plant growing larger and larger. When it's larger than you are, imagine a magic door that opens up. Step into the plant.

Take a deep breath. Turn to the east inside your plant, and notice what you see and hear and feel and sense.

Now take another deep breath. Turn to the south inside your plant, and notice what you see and hear and feel and sense.

Now take another deep breath. Turn to the west inside your plant, and notice what you see and hear and feel and sense.

Now take another deep breath. Turn to the north inside your plant, and notice what you see and hear and feel and sense.

Now take another deep breath. Turn to the center inside your plant, and notice what you see and hear and feel and sense.

Take some time to explore the world inside your plant. You may see something that looks and feels like the physical plant, or you may see or sense images, hear sounds, feel emotions, or notice smells that are associated with your plant.

Sit inside it for a moment, and ask its deva (or guiding spirit) to make itself known and speak to you. Ask the plant if it has information for you. Take the time you need to listen and to learn.

Ask if there is a way you can call back the energies and powers of this plant when you need them, an anchor, a word or phrase, an image.

Ask if there is an offering you can make or some way you can give back to the plant spirit.

When you are done, thank the spirit of the plant, and all beings you've encountered.

Turn to the center and say goodbye and thanks.

Turn to the north and say goodbye and thanks.

Turn to the west and say goodbye and thanks.

Turn to the south and say goodbye and thanks.

Turn to the east and say goodbye and thanks.

Remembering your anchor, say goodbye and thanks to the plant, and find the magic door. Walk back out, closing the door behind you. See and feel the plant become smaller and smaller, until it is back to its normal size.

Open your eyes. Breathe, stretch, say your name out loud, and clap your hands three times.

Thank all you've invoked, and open the circle.

Once you have a plant as an ally, it will tell you what you need to know about it and how to use it. You can have more than one plant ally.

You can also use a similar technique to connect with the deva of your whole garden, or of a forest, or of a section of your land. Before I dug swales and planted olives on one part of my land, I spent a long time making alliance with it, asking its advice and permission, making sure it felt okay about the changes I was going to inflict. We carefully spared patches of native vegetation and left a giant coyote bush in the center, as native habitat and to attract beneficial insects. So far, the olives have thrived under my usual regime of benign neglect. The spirit of the land seems to have welcomed them.

BLESSING FOR EARTH

We give gratitude to the earth, to the dust of stars that congealed into the body of this planet, our home, and that still gives form and solidity to our bones and flesh. We honor the rocks, our sisters and brothers, and their long, slow cycles of transformation into life and back to seabed, mountain, stone. We give thanks to the living soil, the mother's flesh, and the billion creatures that haunt her caves and pores and chasms, to the beetles and the ants and the termites, to the soil bacteria swimming in the slick of water that clings to her mineral archways, to the worms, wriggling, eating, coupling, and transforming within her. We bless the plants, the roots and stems and boughs, the great trees reaching upward and the deep-rooted herbs pushing down, all who contribute to the cycles of birth and growth and death and decay that lead to fertility and new growth. For all that feeds and sustains life, for all that grows, runs, leaps, and flies, we give thanks. Blessed be the earth.

ELEVEN

The Center

The Sacred Pattern

From my journal:

I am flying over the Southwest on a clear and beautiful day. For once, I have a seat in the front of the plane and can see more than a wing. Below, the hills ripple and undulate in tones of gold and sand. They are cut by the lines of streams and rivers that carve out their canyons and valleys.

I'm looking at the land, the patterns of light and shadow, noticing how the little streams flow into bigger streams in a branching pattern. There is a regularity to the shapes of those branches, just as there is a regularity to the pattern and size and number of veins in a leaf. As if I were looking down at the skeletons of invisible leaves laid over the land, each ridge with its streams seems cut in regular intervals, each larger river fed by so many streams.

And—it's hard to describe this, but suddenly I can feel the rhythm of the land, as if the streams and rivers were notations for a drum rhythm, marking off space in exactly the same way the beats of a rhythm mark off regular intervals of time. I could play the ridges and the rivulets on my drum, and I understand that those rhythms, those relationships and ratios, repeat themselves in some way throughout nature. I don't know much about sacred geometry, except that it exists, but I'm sure that this insight is what Pythagoras and Co. were all about. The earth is alive, and there is a structure to her body, a unifying heartbeat expressed in the shape of the land and the patterns of flow just as our veins and arteries are related to our heartbeat and carry our pulse. And our veins reflect the same pattern that I see before me on the land.

Pattern and Relationship

We've taken a journey around the circle of the elements. We've been introduced to each of them in turn and have learned something about the cycles of life that each represents. But the elements do not exist in isolation. They are always in relationship to one another.

Symbolically, the center is the point where the elements connect and transform, where that ethereal fifth sacred thing we call "spirit" arises. To understand how the elements of life interact, we need to look not at the isolated elements, but at the patterns around us. Magical consciousness is pattern-thinking, thinking that can comprehend not just separate parts, but wholes in relation to other wholes.

Much scientific and academic thinking is like a focused laser, beaming intensely at one aspect of a subject. Such vision might focus in detail on a needle of a redwood tree, but it could never show us the whole tree itself, let alone the forest. A thousand directed beams might reveal a thousand needles or patches of bark or single buds, but to see the tree we need to step back, broaden our view, and look in a different way. And in fact the most brilliant and creative scientists do just that. Science is full of great discoveries that came in a dream, or to a person lying under a tree, in a moment of relaxation when the laser beam was turned off and a more diffused gaze could take in the sunlight and the stars.

Those who are wedded to their laser beams defend their preferred form of perception as "objective" and "real." Narrower academics will often attack someone whose perception is grounded in the larger patterns. As I've been working on this book, I've also been involved in making a documentary about the life of Marija Gimbutas, the archaeologist who did so much groundbreaking work on the early European Goddess-centered cultures. I've become aware of the tremendous backlash against her work in academia. There are many reasons for the backlash—to assert that women once held power, that war is not inevitable, and that early cultures were based on cooperation, not violence and competition, threatens some of our most basic assumptions about human nature and culture. But part of that backlash also comes because Marija was a pattern-thinker. When Marija was doing her later work, archaeology as a discipline had turned away from attempts to interpret or find meaning in ancient artifacts, and had restricted itself to description. Much of the criticism of her work takes an isolated object from ancient Europe, examines it in a laser-focused lens, and concludes that one cannot "prove" that it represents a Goddess or anything else.

But Marija was not looking at isolated objects; she was looking at patterns. She had an incredible breadth of knowledge and experience. She read fifteen languages and had spent decades examining archaeological reports from all over Europe, including many from eastern Europe that most of her colleagues in the U.S. could not read. She examined thousands of artifacts in museums, in the field, and in the five major excavations she directed. Her interpretations came from the patterns that emerged as themes and images and forms repeated themselves again and again. The evidence she amassed was of a different kind than that of her colleagues, one that did not fit the narrow definitions of the academics.[1]

A pattern is a form or a set of actions, a way of organizing energies, that repeats. Nature is full of patterns—indeed, she seems to enjoy certain patterns and uses them over and over again. Understanding those patterns will help us to a deeper appreciation of how things work in the natural world, and how these patterns impact human culture.

And we use these patterns in our energetic and ritual work, as well. Included below are suggestions for using the basic patterns of nature in ritual and spellwork. You might draw them, paint them, bring in objects that embody them for an altar. You might include actions in a ritual based on one of the patterns, or build a ritual around them. You might simply visualize them, knowing that patterns shift and channel energies, and energy generates form.

When we can observe and truly understand some of the basic patterns in nature, we can learn not just to speak back to her, but to sing with the music of the spheres.

GENERAL PATTERN OBSERVATION

In your home base, ground and come into your senses. Take some time to observe patterns. What patterns do you see in the earth? The water? The living things around you? What forms repeat?

If you are doing this alone, you might want to take a sketchbook with you. Drawing the patterns will help you experience them in a kinesthetic way.

If you are doing this exercise in a group, have people observe on their own and come back and draw the patterns they saw on butcher paper or a blackboard.

Then share your patterns. Did several of you notice the same pattern? Do the patterns fall into groups? Why do you think they repeat? What do these patterns do?

Patterns direct and channel various energies. They repeat because they *do* something. As we go more deeply into specific patterns, each section will begin with an observation exercise. I strongly recommend that you do the observation before going on to read the rest of the section. You can cheat, of course, but here's what will happen if you do: your vision will already be shaped by expectations, and your observations will not be as open and fresh as they would be if you observed first.

Branching Patterns

BRANCHING PATTERN OBSERVATION

In your home base, ground and come into your sense. Begin to observe the branching patterns around you. Are there trees? Other plants? People or animals? Look at a tree and count the number of "orders" (or levels) of branching, from leaf to twig to branch to limb to trunk. How many times does the tree branch? Observe different trees. Is there a change in thickness at each order? What angles do the branches make?

Branching is one of nature's core patterns. Branches are patterns of flow, of collection, concentration, and dispersal. They reflect the physics of water, since most often what living organisms collect and disperse is carried in a fluid medium. And they are two-way flow patterns. Each rootlet of a tree draws up water, which is carried to a large root, up into the trunk, and out through branches and twigs to the leaves; each leaf collects energy from sunlight that is used to make sugars, which travel the reverse route down to the roots. Every root or leaf has a direct route to the central trunk—but two leaves on opposite sides of the tree have no direct link with each other.

Branching patterns have orders of branching, from capillary to vein, from twig to branch. Rarely do they include more than seven orders of branching, however—whether they are river systems or oak trees. A seven-order oak is probably an ancient and venerable tree.

The branches themselves increase in size as we move up the orders, and this, too, is relatively consistent across different systems. The ratio of your capillaries to your veins is about the same as the ratio of a twig to its next-order branch—about one to three. Why these things should be similar is somewhat of a mystery, but it is likely that it has something to do with the physics of moving water.

Branches are usually set at an angle that facilitates movement. Moving bodies lose energy if they stop, make sharp turns, or double back on themselves. Most hardwood trees branch at a roughly thirty-degree angle. Conifers are the exception: their branches head out at a ninety-degree angle from the trunk (although gravity eventually pulls them down). Highway planners apply the same principle when designing freeway onramps and cloverleafs.

Branching patterns are also the model for many human organizations, from corporations to the military. They are the model of hierarchies, of systems that collect something—generally labor and product—from a broad base, concentrate it and disperse it, and return something else, perhaps pay and directions. The branching pattern is generally thought of as embodying efficiency, and for many things it is indeed very useful.

In a tree, the hierarchy of twig to branch to trunk does not reflect a hierarchy of value; all parts of the tree are vital to its functioning. The cells of the trunk don't sit around congratulating themselves on their superiority, constructing ideologies that explain why they deserve that concentration of sap, or putting down rebellions of the disgruntled root cells ready to throw off their oppression and rise up from the underground.

But in human systems, hierarchies often reflect an unequal distribution of power and value, which quickly becomes an unequal distribution of wealth and well-being. Hierarchies can be useful—in human as well as natural systems. I certainly wouldn't want to try to persuade a tree to grow in some other form. But when hierarchies become unjust, when some members are relegated to positions of low status and value and others elevated, we need to find other forms of organizing.

Understanding how branching patterns work can give us insight into how to take them down when they become dysfunctional. Getting those lowly rootlets to communicate with each other, to trade energy and goods directly instead of sending them up the pipeline, is the time-honored way of challenging an unjust hierarchy. Organizing the workers, the students, the ordinary voters—those who form the base of the tree and provide the nutrients—reminding them that they ultimately hold the power to determine the tree's survival, begins to transform this pattern into something more equitable.

If we are designing something that involves movement and flow, perhaps a path through the garden or a drip irrigation system, we need to consider the lessons of the branching pattern. We need to think about the angles where paths join, and shape them to facilitate movement. We can't have too many orders, and we need to widen the arterial paths.

The tree of life is a powerful symbol in many traditions. The early Norse saw it as Ygdrasil, the world tree, a huge ash with a cave at its roots where the Norns, the Fates, dwelled. High in its branches lived an eagle, and a squirrel ran up and down the trunk, communicating between earth and heaven.

Goddesses and Gods were often associated with sacred trees. Athena was identified with the olive tree; Zeus, with the oak. Inanna, in ancient Sumer, was originally the Goddess of the date palm. Asherah, ancient Goddess of the Canaanites, was worshiped in sacred groves and symbolized by a pillar which actually stood in the temple of Solomon for much of its existence.

In ritual, we use the pattern of the tree for grounding, putting our roots into the earth to disperse excess energy, and we use it to gather vitality, concentrating earth energies into the "trunkline" that moves up our spine, extending an energy field out from the top of our head to catch the vital energies of the sun, moon, and stars. The two-way flow of the branching pattern lets us draw earth energies up and move sky energies down, simultaneously. When we are rooted and grounded in the earth, any energetic lightning bolts we encounter will sink harmlessly into the earth and not burn us out. And we are linked to a source of virtually endless vitality.

Circle/Sphere Patterns

CIRCLE/SPHERE PATTERN OBSERVATION

In your home base, breathe, ground, and come into your senses. Now look around you, and begin to observe circles and spheres. Where do you find round forms? How did they get that way? What function do they serve?

The circle is the symbol of wholeness, of completion, of equality, with no head or tail, no top or bottom. As a pure form, a circle encloses more area, and a sphere more volume, with less surface than any other form. That makes them protective shapes. Eggs and seeds, which risk losing vitality if they are exposed to too much heat or cold or other impacts from their environment, are often roughly spherical, to expose less surface area to weather and rain. Pebbles in streams and the ocean, rolled and tumbled and bashed around until all their edges are smoothed, eventually take on a spherical form.

In *Metapatterns*, Tyler Volk writes, "Whenever life needs to close, contain and separate by minimizing area of contact with the environment, a sphere is often the answer. Because the emphasis is on protected, internal development, animal eggs, seeds and buds of flowers and leaves do well as spheres."[2]

When we want to symbolize equality in a group, we sit in a circle. In a circle, each person can see every other person. We cast a circle to symbolize protection, and do our rituals in circles, which contain and focus the energy we raise. If we are drawing or writing out an intention or a spell, we often put a circle around it to symbolize its completion.

Spiral Patterns

SPIRAL PATTERN OBSERVATION

In your home base, breathe, ground, and come into your senses. Now look around you for spiral forms. You may find snail shells or, if you're lucky, coiled snakes. But look also at the trees. Do some of them wind upward like an elongated barber pole? Stand at their base, and look up into the branches. Is there a spiral pattern to the way they attach to the trunk? Look down into the head of a sunflower, or look at the way peas reach out for support. Where else do you see spirals?

A spiral is a dynamic form of a circle. It comes back on itself, but always with a difference. It moves somewhere. The spiral is a pattern of growth and an ancient symbol of regeneration and renewal.

The nautilus spiral is a perfect exemplar of one of the classic forms of sacred geometry. It embodies what is called a Fibonacci series, a term for a mathematical formula that says, "What is, plus what was, is what will be."

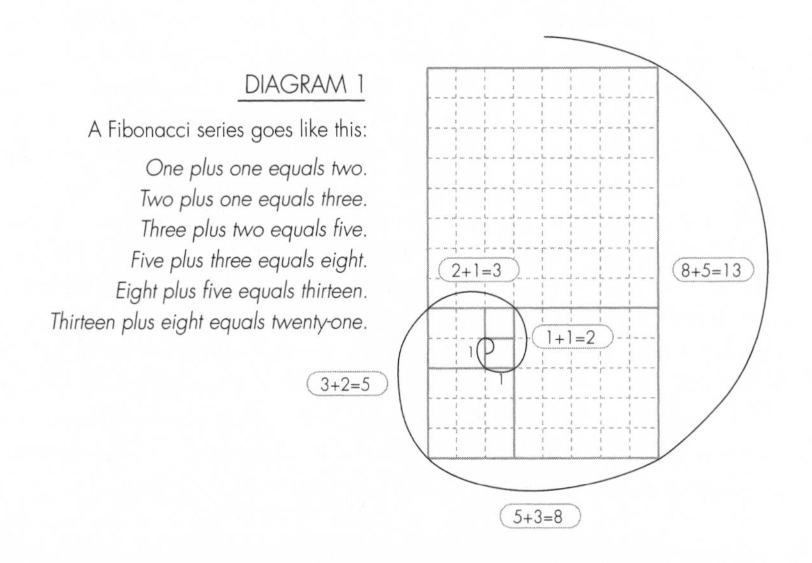

Draw this out, as in Diagram 1, and you have a nautilus spiral. The ratios of the numbers within each Fibonacci pairing give the golden mean, the basis of the sacred architecture of the Greeks. The Parthenon is the best-known example. Spaces based on these ratios feel intuitively right to our bodies. In fact, our bodies embody a rough Fibonacci series, in the ratios of finger joints to finger, fingers to hands, hands to arms, etc.

Tree trunks embody a different spiral, the elongated barber pole. The spiral twist gives strength to the trunk. The elongated spiral is also the form of the drill, a form for penetrating. Some seed pods take this form, drilling into your socks (or elsewhere) to implant themselves in a new environment. Water moves in spirals. A flowing river is an elongated vortex, drilling its banks, as Alice Outwater notes,

> Water that flows in a river moves like a corkscrew, twisting in on itself. The water on the river bottom is slowed by friction as it moves over obstructions on the riverbed, while the water on the surface of the river flows more quickly. Where the river bends, the faster flow at the surface pushes against the bends' outer bank and erodes it, while the slower, siltier water slips to the inner bank and drops some of its load of sand or gravel. This process creates the meanders found on lowland rivers.[3]

A spiral can stretch space. One classic permaculture design is the herb spiral, a round bed built up into a three-dimensional spiral planting surface, which adds planting space and creates interesting edges and small variations in light, shade, and temperature to please different plants.[4]

Time is a spiral, cycling back on itself but at the same time moving ahead. In the spiral dance, we coil in on ourselves, then turn outward to face each person in the group as we pass. When we wind the spiral in, we concentrate energy, eventually releasing it as an upward-spiraling cone of power.

In the northern hemisphere, the sun moves clockwise across the sky, and water forms a clockwise vortex when it drains down a hole. When we raise power for a positive end, to draw in energies and resources or to create something, we move in a clockwise or sunwise direction—or *deosil*, to use the old term. When we want to release or undo something, we move *widdershins*, or counterclockwise. In the southern hemisphere, the sun and whirlpools move counterclockwise, and our magical directions are reversed as well.

The spiral or vortex is a powerful magical form. It's the brew stirring in the cauldron. I might spin a vortex of energy counterclockwise to release obstacles to a project or plan, then spin it clockwise to draw in resources and power. I work with the form of the spiral whenever I want to create transformation and regeneration.

The spiral is one of the most widespread and long-lasting sacred symbols. Spirals are carved on the megaliths of Newgrange in Ireland, dating to the third millenium B.C.E., and decorate pottery and sculpture throughout Old Europe. The snake, which sleeps coiled in spirals and sheds its skin to emerge renewed, was an ancient symbol of the Goddess in Europe. Transformed into the dragon, it represented luck and renewal in ancient China. Spirals grace the pottery of the Hopi and the carvings of the Australian aborigines. Throughout the world, the spiral has been recognized as one of the basic patterns of life and growth.

THE SPIRAL DANCE

In ritual, we often dance the spiral. The spiral dance symbolizes regeneration. It allows everyone to come face-to-face with everyone else in the circle, and to look into one another's eyes for a moment, a moving meditation. And it can wind up a vortex of energy, serving as the base for raising a cone of power.

The spiral dance works best with forty to three hundred people, although I have done it successfully with close to two thousand. With fewer people, you will end up with more of a double line than a full spiral, and the dance will not last long enough to build a lot of power. However, you can spiral in and out several times to extend it, and still have a lot of fun.

Start in a circle. To end up dancing *deosil*—that is, clockwise—the leader frees her left hand and begins spiraling inward. When the spiral is about three lines thick, or when the center begins to look small, she turns to face the person following her, and keeps going (see Diagram 2).

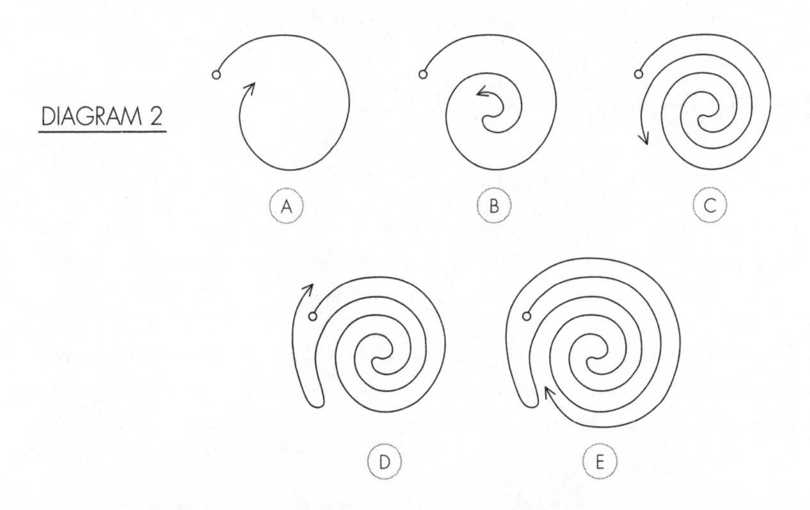

DIAGRAM 2

She continues around the inside of the circle, winding between the lines. Her path gradually takes her (and the line she's leading) out. The line following her will be passing in front of others, and people can look into each other's eyes.

When she passes the tail of the spiral, she should keep going about a third of the way around the circle. Then she again turns to face the person following her. Now she has made a loop in the line, and is moving around the outside of the circle. Again, everyone will have a chance to look into each other's eyes.

When she completes a circle and returns to the loop, she goes inside the loop. If by chance she should encounter the tail of the spiral, she must stay outside the tail. If the line unpeels and the loop disappears, she should hold the form of the spiral in her mind and simply spiral in again.

When she returns to the center, she should try not to let the center get too tight and closed. One trick is to spot a person in the line behind and keep the circle turning at a pace that keeps her even with that person as the back portion of the spiral continues to unwind. This sometimes takes some subtle communication with elbows and butt, as people tend to stop and crowd in when they reach the center, forgetting to be aware of the rest of the spiral behind them.

When the tail has unwound, and the energy is strong, the group can begin to focus on raising a cone of power, an energy vortex centered around an intention embodied in a chant, an image, or an object in the center. After a time, the chant turns into a wordless sound, like an open "Om." Participants can use their hands and bodies to direct energy, generally throwing their hands up into the air at the end and sounding together in a powerful harmonic tone. That's the moment to visualize your magical image and hold your intention strong. The energy will eventually die down, and participants can put their hands on the earth, crouching or lying down, to ground the energy and feed the earth.

The spiral dance does not need to end in a cone of power. It can wind and unwind back into a circle, or stretch into a line or snake dance.

Radial Patterns

RADIAL PATTERN OBSERVATION

In your home base, breathe and come into your senses. Now observe the radial patterns around you. Are there flowers in bloom that exhibit this pattern? Dandelions about to blow? Where else do you find this pattern?

The radial pattern is, on the one hand, the target, the pattern that says, "Come here! Come to this center!" Flowers use it to attract pollinators. Spiders incorporate it into the spokes of their web.

But a radial pattern is also the pattern of an explosion, radiating from the center, dispersing elements equally in every direction. It's the starburst, the sunburst, the seed pod exploding or dandelion blowing to send its seeds out into the world.

If I am doing a ritual or a spell to draw in some positive influence, I might create a radial pattern—draw an image, maybe, or place flowers on my altar.

Radial patterns and sun forms are also ancient sacred symbols, found in the earliest cave art, in pictographs from Norway to the Sahara. Sun Gods and Goddesses abound in mythology: Gods such as Apollo of the Greeks, Ra of the Egyptians, and Goddesses such as Sulis of the Lithuanians, Amaterasu in Japan. The Goddess is also the Star Goddess of Wicca, Inanna, Ishtar, Astarte, Aphrodite, and Venus, all love Goddesses associated with the morning and evening star.

Random/Scatter Pattern

RANDOM/SCATTER PATTERN OBSERVATION

In your home base, ground and come into your senses. Now begin to look for randomness, for what has no fixed pattern. The scattering of leaves on the ground, of stones in a stream. Where is there true randomness, and where do you sense an underlying order?

Randomness in nature is what allows freedom, movement, change. Randomness dissipates energy and breaks down form. A random scattering of rocks below a waterfall will help break the force of the water. A scatter of leaves on the ground will break down into compost.

Many cultures have honored this principle as the Trickster or Fool. The Raven/Trickster of the North American Northwest Coast First Nations was also the Creator, who brought fire to the world. Coyote, in the cultures of the West Coast, was the Trickster, whose practical jokes and bawdy humor helped keep the world in balance. Alegba, in the Yoruba tradition of West Africa, is the Trickster/Communicator, who translates messages between the world of humans and the realm of the ancestors and orishas, the greater powers beyond. The Fool in the Tarot, depicted with his knapsack and dog, about to step off a cliff, represents the freedom to take a leap, to step into the future without knowing what the outcome will be, to take a fall and land on your feet.

When energies are stuck, when forms are too rigid, introducing some randomness can help bring freedom and creativity back into a system.

Packing and Cracking/Honeycomb Pattern

PACKING AND CRACKING/ HONEYCOMB PATTERN OBSERVATION

In your home base, ground and come into your senses. Now begin to look at the different patterns of bark on trees. What are the different shapes and textures, forms and lines? Where else can you see these patterns? Look at your own skin. Observe the cracked mud in a dried-up mud puddle.

The patterns of bark and skin are patterns of expansion and contraction. A tree expands as it grows. The gray, smooth skin of a young Douglas fir grows thicker and deeply grooved as the tree ages. Our own skin expands and contracts, wearing lines into its smooth surface over time. As mud dries, it shrinks, leaving cracks and fissures.

This packing and cracking pattern is also a pattern of drainage. Channels in tree bark collect water, concentrate moisture, and move it off the surface of the tree (where it could cause rot) and down toward the roots. One way farmers in very dry lands conserve moisture is by erasing the cracks that form after rain or irrigation water evaporates, thereby preventing the remaining moisture from being channeled away. Someone who needed to drain marshy or waterlogged land might employ this pattern in digging a netlike system of water channels.

This is also a pattern of packing. Take a bunch of small, soft globes and attempt to push them into the smallest possible space: they will crunch into cells like the cells of a beehive or the geometric seeds of a sunflower. A beehive's six-sided cells are stronger than cubic cells of the same volume would be, and they pack better than spherical cells.

The amazing bee, who constructs her elaborate hives and gathers and stores honey there, who lives in a complex social system and carries out complex tasks, nurtures and feeds her young, and communicates through scent and dance the location of distant food sources, has been a sacred symbol in many cultures. Melissa, the Greek Bee Goddess, would be a fitting patroness of this pattern.

Web Pattern

WEB PATTERN OBSERVATION

In your home base, ground and come into your senses. Now begin to look for webs. Are there spiderwebs around? If so, what shapes do they take? Where do you find them? When are they visible, invisible? How do they fill space? Catch light? Transfer tension and compression? What elements are connected to which other elements? What other patterns do they incorporate? Where else does the pattern of the web repeat?

The web is one pattern in nature that we use as a magical symbol again and again. The spider's ability to spin, weave, and construct her elaborate creations has always fascinated people. It seems to hint at some larger, creative power at work in the universe, spinning reality out of her own body, weaving the tapestry of life. The Weaver, the Spider Goddess, has been sacred to many cultures. In the Southwest, she is Spider Woman of the Pueblo people, who wove the universe out of thought. In ancient Greece, she was Arachne. The Fates of the Greeks spun out a person's lifespan, measured the thread, and cut the cord. The Norns of Scandinavia were also spinners and weavers of fate.

Sitting in my home base spot one morning, I noticed three different kinds of spiderwebs within a few feet of where I sat. The first was a classic web, a two-dimensional spiral overlaid on a radial pattern. The rays of this sort of spiderweb are the anchors, giving it strength and support. Those delicate strands are stronger for their thickness than steel cable. The spiral that winds around them is the sticky trap. Movement at any point on the web sets up a vibration that alerts the spider that something has been caught. These webs are virtually invisible until the sun hits them at just the right angle, whereupon they glow like a brilliant idea, their bright, iridescent design shimmering for a short time and then disappearing.

As I looked out at the world from my home base that morning, I saw that another spider had built a domed web, an awesome engineering feat of suspension and tension that hung in the trees. This sort of web is likewise invisible until the sunlight illuminates its hanging fairy domes and palaces.

And a third web within my field of vision that day seemed almost entirely random, a wild zigzagging of sharp angles that filled space with a postmodern disdain for the usual forms.

The web is also one of the key metaphors of our age, with the emergence of the World Wide Web as a vital form of global connectedness.

If the branching pattern of hierarchy is the underlying form of our current political/economic system, the web is the pattern that seems to be emerging as a counter. Direct democratic organizing creates weblike networks of small circles linked to other small circles through spokescouncils (see Diagram 3). The difference between a web and a tree, in organizational structure, is that in a web all the parts can communicate directly with all other parts, if desired. Centralization can be useful. For example, if you want to know what programs

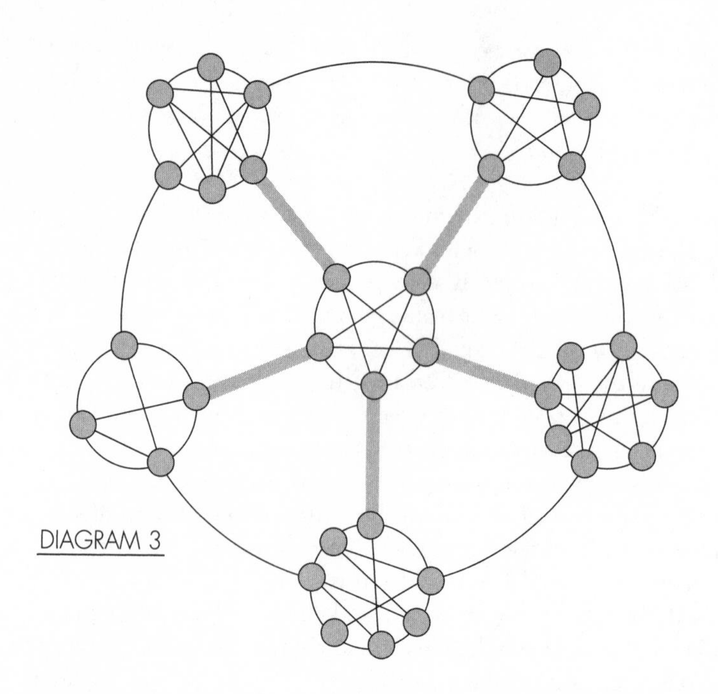

Reclaiming groups are running this summer, it's easier to go to one Web site than to check ten; if you want to find out what peace groups are operating in your area, it's much easier to go to unitedforpeace.org than to surf the Internet or thumb through the phone book or drive around town looking for them. But if the Marin Peace and Justice Center and the Santa Rosa Peace and Justice Center want to collaborate on a forum, they can talk directly to each other about it and make their own decisions. They do not need permission from some central authority.

Groups, spaces, and communities need a center that can further connectedness, a place or time when people gather for unplanned, spontaneous interactions that are often the matrix of new ideas and creativity.[5] In a home, people often gather in the kitchen, or around meals. A village has a marketplace or central square. A gathering or encampment needs a daily circle, or at least a time for announcements. Without a center, a web-based structure may lack a sense of identity and cohesion.

But centralization also means vulnerability. The key Web site can be hacked; the village market, commandeered by tanks. Whenever a structure becomes centralized, it always needs a backup. One of the permaculture principles is redundancy: we should always have more than one way of providing for every function.

WEBS IN RITUAL AND ACTION

Magically, the web is used to symbolize interconnection. Witches have many songs and chants that invoke the web and the weaver. We might sing,

> Weave and spin, weave and spin,
> This is how the work begins.
> Mend and heal, mend and heal,
> Take the dream and make it real.
>
> Or:
>
> Breath by breath, thread by thread,
> Conjure justice, weave our web.

To weave a web in ritual, provide the group with balls of yarn about the size of a tennis ball. Distribute the balls around the circle. All participants hold the end of their thread, or attach it to themselves, then toss the ball across the circle to someone else. As the balls fly back and forth, the web is woven. Often it's useful to have a "spider" or two willing to crawl beneath the web to retrieve balls that fall short. People might call out qualities they wish to weave into the web as they toss their balls.

When the web has been created, the group can circle with it, dance with it, and raise power to charge it. When the ritual is done, carefully take the web apart, perhaps saying something like "As we break this physical symbol of the web we have woven, we ask that the web of connections remain strong and whole. We will each take a thread to keep us linked to this web."

Each person can take a thread of yarn home with them. People can tie the yarn around each other's wrists while making a pledge or naming ways in which they will stay connected.

A web can also be useful in nonviolent direct action. It makes an excellent soft blockade, one that can quickly fill an intersection or block an entrance, delaying the authorities without heightening the level of tension. Chainsaws cannot cut through yarn, so forest defenders have used webs to slow down logging crews. When the yarn is first charged in ritual and then used in an action, it retains great power. I remember one blockade in Washington, D.C., when we were protesting a meeting of the International Monetary Fund and the World Bank. Our affinity group was in an intersection that had been filled with protestors before we even got there with a giant web, using yarn we'd all charged in ritual the night before. The police never tried to clear us away. Late in the day, one of our friends came back from a walk smiling. She had passed by a group of police cars and heard them discussing whether or not to clear our intersection. "No, too much yarn to deal with," was their conclusion.

Patterns for Meditation

There are some patterns that have been used for meditation and insight in many different spiritual traditions. The mandala and the labyrinth are complex patterns that can help us shift our consciousness into pattern-thinking.

The Mandala

The mandala is a quartered circle, a visual representation of the four elements within the circle of the whole. Tibetan Buddhists people their mandalas with deities and a complex pantheon of mythological beings inhabiting the four quarters, making the mandalas maps to the various realms of the spirit world. Psychologist Carl Jung recommended drawing and painting mandalas as a way to integrate one's own psyche.

But perhaps the most elegant, and certainly most vitally important, mandala is the one created by life itself to perform the miracle transformation of photosynthesis: the mandala of chlorophyll.

In *Gaia's Body*, Tyler Volk writes,

> Arguably the most important molecule of the biosphere, chlorophyll would be a perfect icon of a science-based, earth-centered religion. In the form of a molecular model, its head and tail might easily replace, for example, the Catholic chalice. A nature priestess could hold the iconic molecule by its tail and lift its illuminated, green head, truly the bringer of light to life, glittering high in the air before a reverent congregation. That might even lure me back to church![6]

MAKING A CHLOROPHYLL MOLECULE

If we had a Gaian Goddess temple, the chlorophyll molecule would make a lovely stained-glass window or floor mosaic. But, in the spirit of not taking ourselves too seriously, here's a story and directions for making a chlorophyll molecule—an enterprise that can be done with a minimum of twenty-five children or childlike adults:

The group facilitator begins by telling the following story:

Once upon a time, a long time ago . . .

A very young Gaia was amusing herself, watching the first life, the billions of simple, one-celled fermenting bacteria chase each other around her seas, gobbling complex molecules to fuel themselves, promiscuously trading genes. They were increasing at such a rate that she began to worry.

"How will they feed themselves when those complex sugars run out?" she wondered. What would they do? Would they starve? Would the seas empty of the life that she found so wriggly and entertaining?

As she mused, she began idly arranging and rearranging molecules in pretty designs.

Then the facilitator says to the group,

Now you be the molecules.

Someone jump in the center and be magnesium.

Four of you stand around her in the four directions and be molecules of nitrogen. Nitrogens—face out and hold out your hands.

Now four others of you go to each of the four nitrogens and make a little circle of five. You are carbons.

So now we have a center of magnesium, surrounded by a circle or pentacle in each of the four directions, with a nitrogen at its head and four carbons.

Now four more of you go and be a link between each circle. You are also carbons.

And all the rest of you link up and make a long, carbon tail.

DIAGRAM 4

Core Mandala within the Chlorophyll Molecule

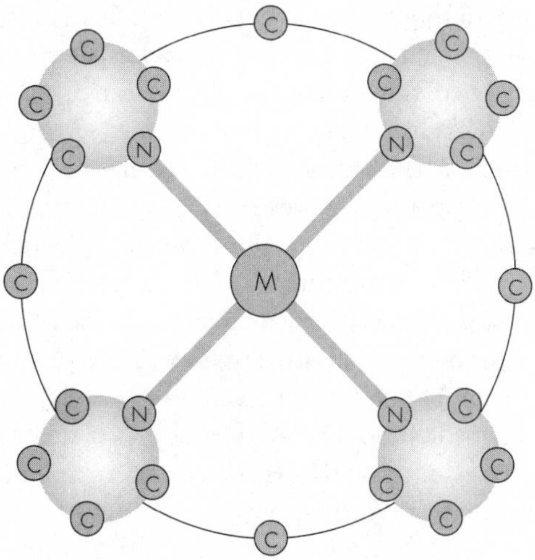

When the group members have carried out these directions (as shown in Diagram 4), the facilitator continues:

And now we've made the chlorophyll molecule, the beautiful mandala that Gaia designed. And when the tail holds the mandala to face the sun, and a photon of light strikes the magnesium in the center, she begins to vibrate. And that vibration is the energy that begins the complex, miraculous process of photosynthesis—the process that created the air that we breathe, that feeds just about all of life today.

In conclusion, the group can sing the chlorophyll song (to the tune of "O Tannenbaum"):

O chlorophyll, o chlorophyll,
You make the green world really real.
O chlorophyll, o chlorophyll,
If we don't love you, no one will.
O chlorophyll, o chlorophyll,
In parsley, sage, in thyme or dill,
O chlorophyll, o chlorophyll,
Make oxygen for lung and gill.

—*Words by Starhawk and Evergreen Erb*

The Labyrinth

The labyrinth is another sacred pattern that is very ancient. Labyrinth designs have been found on standing stones and in pottery designs from third millennium B.C.E. Crete to the contemporary American Southwest tribes. The labyrinth figures in myth: in ancient Crete it was the secret maze below the palace of King Minos where Theseus faced the minotaur, the monster with the head of a bull.

A true labyrinth, however, is not a maze, not a puzzle or trap. A labyrinth has a single pathway through it, not many confusing alternatives and dead-ends, and leads to a center that's a spot for insight and meditation. There are many forms of the labyrinth, from the classic Minoan seven-path labyrinth to the more complex pattern of the labyrinth found in Chartres cathedral. Today, many churches and healing centers are adding labyrinths as places to do a walking meditation. One of the beauties of a labyrinth is seeing many people walking and meditating together, each on her or his own unique journey, but sharing a sacred space in common.

MAKING A LABYRINTH

A labyrinth can be made simply, drawn in chalk on a floor or pavement, laid out with masking tape on a rug, mowed into a lawn or field, or laid in stone, paving, or gravel amidst plantings. For a walkable labyrinth, a space at least twenty feet on each side is needed.

First, before tackling your actual labyrinth medium, *draw* a labyrinth. (Diagram 5 illustrates the Minoan seven-path labyrinth.) Take a sheet of paper and make a cross (A). Put a small, right-angled arrow tip with its point toward the center in each quadrant, and a dot in each corner to make a square (B).

DIAGRAM 5

We can start on either side, but let's start on the top right. Connect the top of the center line to the top of the first arrow tip on the right with a curved loop (C). Now move counterclockwise and connect the top of the left arrow's tip to the top of the first dot on the right (D). Move counterclockwise again, and connect the dot on the left to the tip of the sidebar of the arrow tip on the right (E). Continue around, connecting the elements, until the labyrinth is complete.

What you've drawn above are the *walls* of a labyrinth. When you're comfortable drawing that form, experiment with drawing the path *between* the walls instead (see Diagram 6). When you mow a labyrinth, you mow the path, not the walls.

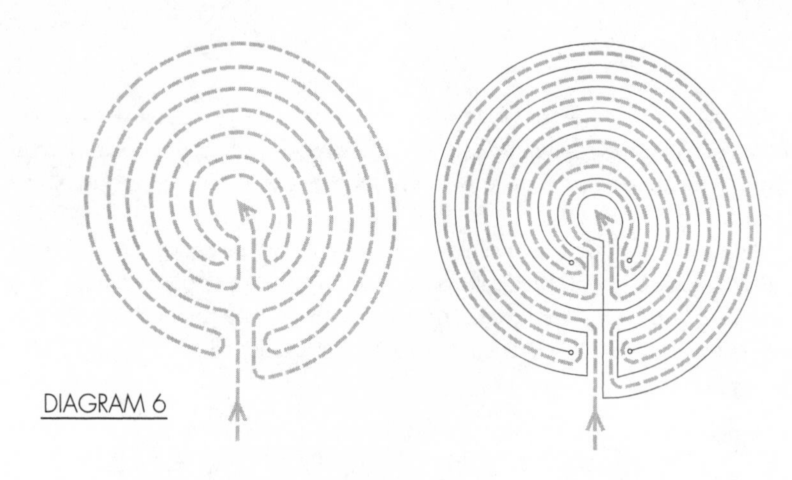

Practice chalking labyrinths or laying them out with straw or tape or some impermanent material. When you are comfortable with the form, you can easily begin to create more permanent labyrinths in the landscape.

There are many ways to use the labyrinth as a meditation tool.[7] For me, perhaps the simplest is to hold an issue, problem, or question in my mind and walk into the labyrinth slowly, examining the issue from all angles, changing my focus each time I switch directions. I might ask myself, for example, how I think or feel about it, what material factors are involved, what the heart of the matter is, what spiritual energies I can call on to help me, what my vision is, and how it might manifest.

When I reach the center, I stand in open, receptive attention, noticing what insights or visions come.

As I walk out, I focus on how I might use those insights or manifest those visions.

Before entering and on exiting, I leave an offering of waters of the world.

My favorite labyrinth is on the headlands above a beach in Sonoma County. It's a simple stone labyrinth laid on the low scrub, with a view beyond of the waves and cliffs. A path is worn into the ground from the many feet that have walked it, and in the center is a simple altar of stones, which is always graced by many offerings. The rumor is that a local woman and her friends built it to celebrate her fiftieth birthday. It's not on any map, but many people have discovered it. I stop by often when I need inspiration or grounding, and I especially love the sense that I am sharing this sacred place with many unknown others. The offerings are always changing, and sometimes stones are added to the path or taken away, but the labyrinth remains, a beautiful gift of the spirit.

Patterns in Our Lives

"Why do I always fall in love with men who abuse me?" "Why do I always do something to undermine myself just when success is within reach?" Much of emotional healing work and psychotherapy revolves around identifying and changing our own internal patterns, especially the destructive and dysfunctional actions and relationships that we tend to repeat.

We can apply the skills of observation and awareness to our own patterns as well as to the patterns of nature around us. Awareness is the first step to insight, and insight is the first step to change.

INNER PATTERN OBSERVATION

The magic circle of the elements can help us look for our own internal patterns. You can work with the questions below in several different ways. Alone, you can meditate on them, write in your journal about them, or take each for a period of days or a week to reflect upon and keep notes on.

With a friend, partner, or in a healing or mentoring relationship, you could share with each other your reflections and give each other feedback and insights.

In a group or circle with a high degree of intimacy or trust, you could go around the circle, giving each person time to respond to each question without being interrupted or challenged. At the end, you could discuss what is similar or different about your conclusions. In group work, I would suggest taking one element each session, and devoting five or six sessions to this work.

Air

Are there patterns you can identify in your thoughts? Particular phrases or words you say to yourself again and again, or snatches of inner dialogue that repeat? Names you call yourself? Images you hold? Fantasies?

How do these affect your perception of yourself? Others? How do they impact your energy? Emotions? Body? Spirit? The choices you make? The possibilities you see around you? How do these patterns restrict or harm you? How do they serve you? What do they do?

Fire

Are there patterns you can identify around your energy level and how you use your energies? Cycles or repeating ebbs and flows? Ways you dissipate or squander energy? Patterns of eating or drinking or sleeping? Repetitive ways you build up energy or stoke your fires?

How do these affect your perception of yourself? Others? How do they impact your energy? Emotions? Body? Spirit? The choices you make? The possibilities you see around you? How do these patterns restrict or harm you? How do they serve you? What do they do?

Water

Are there emotional patterns that you can identify? Cycles of feeling? Patterns in love, patterns in relationships? Patterns in the way you respond to fear? Hope? Anger? Attack? Loss?

How do these affect your perception of yourself? Others? How do they impact your energy? Emotions? Body? Spirit? The choices you make? The possibilities you see around you? How do these patterns restrict or harm you? How do they serve you? What do they do?

Earth

Are there physical patterns you can identify? Patterns in your health, fitness, muscle tone, and flexibility? Do you get sick in response to other patterns? Cyclically?

Are there patterns you follow around money or other material resources? Around providing for yourself and others? Around shopping or spending?

Do you notice patterns around your ability to set boundaries, or your encounters with others' boundaries?

How do these affect your perception of yourself? Others? How do they impact your energy? Emotions? Body? Spirit? The choices you make? The possibilities you see around you? How do these patterns restrict or harm you? How do they serve you? What do they do?

Spirit

Are there spiritual patterns that you can identify? Patterns around consciousness-change, intoxication? Addictions? Patterns in communication and connection with others?

How do these affect your perception of yourself? Others? How do they impact your energy? Emotions? Body? Spirit? The choices you make? The possibilities you see around you? How do these patterns restrict or harm you? How do they serve you? What do they do?

GROUP PATTERN OBSERVATION

Groups, too, fall into patterns. Below are some questions about the group's own processes that you might wish to consider.

Are there repeating patterns that emerge around the group's perception of itself or others? Around who has influence, whose voice is listened to by others? Around the group's energy level, the distribution of work and responsibility? Around who holds power? Patterns of attack or judgment? Emotional patterns? Patterns around acquiring, using, and distributing money or other material resources? Setting boundaries and encountering boundaries? Patterns of inclusion or exclusion? Patterns in communication, in connection or lack of connection?

Changing Patterns

A pattern may serve us or restrict us, further our growth and development or truncate it. But if a pattern is not moving us forward, we can change it.

Changing patterns is not easy, however, especially if they are ingrained or rigid, as addictions are. To change a pattern, we may not be able to simply *stop* it; instead, we may need to *replace* it with a new, consciously chosen, pattern. Twelve-step programs such as Alcoholics Anonymous replace the patterns of addiction with meetings that themselves follow a set pattern, and with a program of support and self-reflection. Many forms of counseling and psychotherapy exist that can be very helpful in identifying and changing personal patterns.

Ritual can also be an important tool for changing patterns. Ritual itself follows a pattern, and we can use our understanding of patterns (and what they do) to help us construct a ritual or spell that can move us from insight toward change.

One common pattern in ritual could be called releasing/calling in. Within the body of the ritual, we go through four steps:

We consciously acknowledge and name some pattern, force, or energy that we want to release or transform.

We do something symbolically to release it.

We identify some positive pattern, force, or energy we want to replace it with.

We do some act to symbolize that replacement.

The pattern we are releasing might be negative, but it might also simply be something that has outgrown its current form and needs to change in order to grow.

So to work a changing-pattern ritual, we must first identify the pattern we want to replace and find some object or action that will symbolize that pattern. Let's look at an example.

At one point in the growth of Reclaiming, our tradition of the Craft, we had seven or eight different Witch camps around the United States and Canada, and we called our first-ever retreat for teachers and organizers. At that time, all the decisions about teaching and programming were made by the San Francisco group that had started the camps, and we wanted to broaden the structure and empower more groups. Those of us in that San Francisco group spent many hours in meetings, and I became more and more aware of how many threads of the organization I carried personally. I taught at all the camps, and I knew everyone, all the teachers and organizers. They all knew me, but many had not met each other before that weekend. We wanted to create a web of autonomous camps, but I was still the central hub, and I wanted out of that position.

As part of our closing ritual at that retreat, I stepped forward and said that I felt as if all the energetic threads of all the camps were held in my belly. I said that I felt that I had held them well and nurtured the camps into growth, and that I was ready to let go of them. The image of myself as a spider with all the threads of the camp extending from my body symbolized the pattern we wanted to transform.

I then asked people to come forward and pull the threads out of my belly. As they did so, I felt a tremendous sense of release and relief and lightness, as if letting go of a tremendous burden. Patti Martin, one of our teachers, then reminded me that it was important for me to fill up with some positive energy to replace what I had let go of. I consciously breathed in energy from the earth, and opened to the love and support of the community around me. And, in fact, letting go of some of the work of the Witch camps opened up space and time for me to learn more about permaculture, spend more time in my garden, and (later) focus more on activism and global issues of earth-healing.

To symbolize the new pattern we wanted to create, we wove a web of yarn that represented a more open field of connection. Other challenges arose, both in the ritual and afterward, but the basic transition was made.

RITUAL FOR CHANGING PATTERNS

Before you begin, decide how you want to symbolize the pattern that you want to change as well as the new pattern you hope to adopt, and track down the necessary supplies.

There are many ways to symbolize change. You could find an object that represents the old pattern and draw on it, burn it, dissolve it, or bury it and let it compost. You could draw, sculpt, or paint a new symbol or image for the new pattern, or you could plant a seed, an herb, or a tree, weave or spin or sew a garment, dance a dance, or simply find something appropriate and breathe and pour energy into it. The discussion above of pattern can give you ideas of what to draw or how to symbolize the change.

Once you've decided on your symbols, ground and create sacred space.

Create your symbol for the old pattern. Raise energy with a dance, chant, or action to release it.

Create your symbol for the new pattern and charge it with energy by chanting, dancing, breathing, making sounds, or symbolic action. Ground the energy into the earth, and keep your symbol on your altar or in a special place where you will see it often.

After enacting your transformation, you might offer thanks, bless food and drink, and then say goodbye and thanks to any powers or energies you have invoked.

The power of the ritual is not just in the actions or words, but in all the work you do to prepare for, plan, and carry out the ritual; in the strength of your attention and focus; and in the witnessing of your community. A ritual might not instantly bring about the change you desire, but it might start you on the road to that transformation.

Patterns of Time

The Cycle of the Moon

MOON CYCLE OBSERVATION

For a full cycle, try to actually see the moon each night, even if only for a brief time. Notice when it rises and where, or when it appears in the sky. What time do you have to get up in the night to catch the waning moon? How early does the waxing moon set? Do your energies change with the moon?

In Wicca, our core religious imagery centers around the cycles of the moon and sun, which are two mythological representations of the great cycles of birth, growth, death, and regeneration throughout nature. Because I and others have written so extensively about these cycles, I will be fairly brief here, but much material can be found in *The Spiral Dance* and *Circle Round*.[8]

The moon cycle, from dark to waxing crescent to full, then to waning crescent and back to dark, swells the ocean's tides, awakens the powers of growth in plants, and strengthens human and magical energies. Farmers plant by the moon because they know that seeds will sprout and take root best when they can draw on the energies of growth and increase that the waxing moon provides. Spells and rituals for increase and growth should also be done while the moon is growing. Symbolically, the new moon is the Maiden, the young Goddess, wild and free, infused with the energies of inspiration and beginning.

The full moon is the culmination of power, the flood tide, the peak of magical and subtle energies. It's the time for rituals of fulfillment, empowerment, and transformation. The full moon is Mother, not just of children but of all creative enterprises, the nurturer, she who sustains us. And she is Lover, the erotic Goddess, riding on the flood of sexual energies that are aroused with the moon.

And finally, the waning moon is the transformation of power, the turning inward of the energies of growth, a time for letting go, releasing, going within. The waning moon is the time to plant root vegetables and those that store their nutrients underground. It is the time for divination and meditation, for looking within. Mythically, the waning moon is the Crone, the wise woman, the elder who knows the secrets of life.

When we become aware of the moon's cycle, when we craft rituals that follow her ebb and flow and when we are aware of the flux of her energies, we become embedded in the cycles and rhythms of life. Life ceases to be just a linear progression from birth to death and becomes part of a great round of constant regeneration. The moon is our monthly proof that darkness gives way to light, endings to new beginnings.

The Cycle of the Sun: The Wheel of the Year

At one point in my life, I lived a few blocks from the ocean, and I became obsessed by sunsets. Every night, I had to walk down to the beach to watch the sun set. I saw how it moved northward as the year progressed, until instead of setting over the waves it dropped below the hills that cradled Santa Monica Bay. After the summer solstice, it began to move south again, until in midwinter it lit the crests of the waves with red and purple fire.

The cycles of the sun, its daily rising and setting and the yearly cycle of the seasons, are also core cycles of birth, death, and regeneration. In Wicca, we celebrate eight major seasonal holidays, or Sabbats, evenly spaced around the year, which constitute the Wheel of the Year. They come from the Celtic traditions and are attuned to the climate cycles of the British Isles and similar temperate regions, but because they are evenly spaced, they are adaptable to many variations. Each holiday begins with sunset, not sunrise, so they span two days.

These celebrations include the solstices, the shortest and longest days of the year, and the equinoxes, the two days (signaling the beginning of spring and fall) when day and night are equal. The other holidays are the cross-quarter dates, which fall halfway in between the solstices and equinoxes. These cross-quarter dates were the key markers of the ancient Celtic year.

The names for these eight holidays, listed below, derive from Celtic or archaic English names, and have become traditional in many branches of the Craft.

Samhain, or Hallowe'en, October 31 / November 1

Yule, or Winter Solstice, December 20–23
(The exact dates of the solstices and equinoxes vary from year to year and thus must be checked with an astrological calendar.)

Brigid, February 1/2

Eostar, or Spring Equinox, March 20–23

Beltane, May Eve / Mayday, April 30 / May 1

Litha, or Summer Solstice, June 20–23

Lughnasad or Lammas, July 31 / August 1

Mabon, or Fall Equinox, September 20–23

The following text offers one way to view the Wheel of the Year as an ongoing mythology that ties into our personal growth and development. In this version the forces of change are personified and gendered female and male, but versions could be made that are very different, using multigendered or nonhuman imagery.

The year begins at *Samhain* (pronounced *sow* [as in female pig]-in). Although we are moving into the darkest and coldest time of year at this holiday, we celebrate Samhain as our New Year. The old crops are gathered in; the land is resting, waiting for new growth to begin. The Goddess is the Crone, the old wise one, she who receives the dead and comforts them. The God is the Horned God, the animal who gives away his life so that others can live. The veil is thin between the worlds of the living and the dead at this time, and we can visit our beloved dead and receive help from the ancestors. Theirs are the loving arms that gather us in when life is done, cradle us, and bring us back to rebirth. We honor our ancestors and our beloved dead.

At *Yule*, the Winter Solstice, the year is reborn. The sun, who has grown old and tired, goes to sleep in the arms of Mother Night, and is reborn at dawn. The Goddess is the Dark Mother, the giver of gifts and the teacher of lessons. Her love is the first gift given to all of her children. The God is the reborn year, all that is new, growing and possible.

The year grows, and the sun gets stronger. The days begin to lengthen. At *Brigid*, the year is beginning to grow up. The teacher of lessons gives us challenges and receives our pledges. She is Brigid, Goddess of Fire and Water, the ancient Goddess of poetry, the forge, and healing. The God becomes the poet, magician, teacher—the keeper of the mysteries. We honor all teachers, stepparents, foster parents, aunties and uncles who teach and care for us.

At *Eostar*, the Spring Equinox, day and night are equal and balanced. The Mother steps back, and the Daughter comes forth. She is life itself, who has been sleeping in the dark of winter. Now the sun wakes her up, as seeds awaken and growth begins in the warmth and rains of spring. If the Winter Solstice is the birthday of the sun, the Spring Equinox is the birthday of the earth. The Goddess is the Maiden who returns, bringing Spring, and she is the magic hare who lays the egg of life. The God is the waxing sun, and the Trickster, power of change, freedom and chance, B'rer Rabbit, Raven, Coyote.

At *Beltane*, the earth is fully awake and everything is blossoming. Just as Samhain was a time to connect with the dead, Beltane is the holiday that celebrates life. The Goddess and God become the lovers of all living things, and bless all forms of love. The gates between the worlds are open, and we can connect with the life-spirits of plants, animals, fairies, and the Mysterious Ones. We honor all mothers who bring life into the world.

At *Summer Solstice*, the sun reaches its peak and begins to decline. The Goddess becomes the lover of all things that fade and die. She is the mother of abundance and fruitfulness, who loves and feeds her children. The God, at the height of his power, begins to transform, to go into the Otherworld, carrying our messages and hopes. He represents all beings who sacrifice, who give of themselves so that life may go on. We honor fathers, who plant the seeds of life.

At *Lughnasad* (pronounced *Loo*-na-sa), the days are already growing shorter, although we are at the warmest time of year. The Goddess becomes the Harvest Mother, as we gather in the first of the crops. The God gives us the gifts of the Otherworld: life in the form of the food we eat and the skills and arts of human life. She is the Gatherer; he is the grain cut down to be planted and grow again. He is Lugh (pronounced Loo) of the Long Hand, the Shining-Faced One, and he is Lugh the Many-Skilled, God of the arts and knowledge that allow human beings to live together. We celebrate his wake at this time of hope and fear. We honor all teachers who share their arts and skills.

At *Mabon*, the Fall Equinox, day and night are equal again. We give thanks for all the gifts we have received, for everything we have harvested. The earth begins to prepare for her winter's sleep; the Goddess and God grow old and wise. They come to us in our dreams and guide us into the Otherworld.

To root our practice of ritual in the reality of the earth and her cycles in the particular place where we live, we need to do our own observation and adapt the myth to our own climate, plant and animal communities, and ecosystems.

SUN CYCLE OBSERVATION

Start a journal for a year of seasonal observations. On each Sabbat, note what the weather is like, what is growing or dying, what leaves are turning or falling, what needs to be done in the garden, what bulbs are blooming, what is budding, sprouting, blossoming, ripening, overripe, rotting, rutting, mating, birthing.

Now write your own Wheel of the Year, your own myth for the cycle of seasons as they unfold where you live.

Better yet, do this for three years, or five, or fifty, letting your own myth evolve.

This is the Wheel of the Year that I wrote for my home in the Cazadero Hills.

Wheel for the Coastal Hills

At Samhain, the year begins as the autumn rains return and renew the land. Deer are rutting, apples are ripe, and the year is getting cold and dark. The sun is low, but the streams are not yet full enough to use our hydro system, and we are reduced to running our generators on fossil fuels. We celebrate the rain's return, praying for a year of enough rain, falling gently and spaced evenly enough so that we don't have floods.

At Yule, the year is dark and cold but the rains are, with luck, abundant. The sun begins to return; our micro-hydro generators rev up and light our homes with energy produced by rushing waters. The coyote bush blooms, and so does the rosemary, and the very first narcissus puts out buds to celebrate the rebirth of the year-child.

At Brigid, the land is green, the daffodils are blooming, and the fruit trees begin to blossom. This is pruning time, time to cut away the dead wood and the overgrown branches, and planting time, the tail end of bare-root season, time to take cuttings and plant the last of the perennials and celebrate the birth of lambs and fawns. We plant, now, not just for the year but for the ages.

At Eostar, the land is at the peak of its green beauty. The soil is warming, and bulbs and fruit trees are in high blossom. We can plant peas, lettuce, and arugula, and greet the emergence of the first wild iris down on the coast, and the awakening of the wildflowers.

At Beltane, the wildflowers cover the land in gold and blue. The cool-weather crops have begun to sprout in the garden, and the collards and kale are renewed. Roses are just beginning to open in a good year, and all the land seems alive with erotic, sensual joy. We celebrate our creativity and the joys of all forms of love. We celebrate the year-child now grown into the fullness of power of the lover.

At Summer Solstice, the sun is warm and the warm-weather crops are in the ground and growing. The springs are still full of water, but the rains have stopped and the hills are beginning to turn golden. It is high summer, a time of growth but also, for us, a time of death and dormancy as the land dries up and the year-child passes into the dreamrealm.

At Lammas the land is dry and the springs are beginning to lessen in their flow. We become aware of fire, vigilant to its unexpected appearance. One spark could set this whole land ablaze. We begin to harvest squash and the first tomatoes. The sun is high, lighting our homes, but the streams are dry. We celebrate the fire ritual, honoring fire and asking for protection as we learn again to live with fire's lessons.

At Mabon, the air is cooler. The apples are ripening on the trees, and the harvest is coming in while the sun begins to decline. We ask help to get us through the last, dry portion of the year even as the very first rains begin to fall. The year-child becomes the dreamer, even as we bottle the sweet tastes of summer into jam and hide its fruits away on our shelves.

And the wheel of the year is complete.

BLESSING FOR THE CENTER

We give thanks and gratitude for the center, the heart and the hearth, and for the complex and beautiful patterns of life. We give thanks for the flow of sap through branches, water through rivers, blood through our veins. Thanks for the spiral vine and the sunflower, the daisy and the dandelion, the sunburst and the star. We bless the random element that brings freedom into the pattern, and thank the weaver of the web of life for teaching us that all is interconnected. May we be aware of our patterns, able to change those that do not serve us and to cherish and strengthen those that do. May the cycles of the moon's and the sun's journeys through the year help us remember that light arises from dark, rebirth from death. Blessings on the cycles, the patterns, and the center.

TWELVE

Healing the Earth

I am sitting in a cave in central Mexico, with a group of people who have come together for a week to share Mexican-rooted and European-rooted traditions of earth spirituality. We have made a beautiful *ofrenda*, an altar/offering of seeds and flower petals laid in patterns, to honor Tonantzin, the Aztec earth Goddess, and we have all squeezed into this cave to listen to the voice of the earth. After a time of silence, we sing and chant together, in English and Spanish,

> The earth is our mother, we must take care of her . . .
> *La tierra es nuestra madre, debemose cuidarla . . .* [1]

We raise power, an echoing, resonant tone that rings through the cave. When we ground, we lay our hands on the earth. In my mind, I hear the earth sigh with pleasure, drinking in our energy. "Do this," she says. "Feed me. Tell people to feed me, to consciously feed energy into my energy body. For I am getting weakened, and I need it."

Then suddenly I saw the earth as a great battleground. There were huge forces of destruction like thunderclouds massing, and also enormous forces of love. I resisted the vision, because I don't like to think in terms of dualities and wars of good and evil. But I heard clearly, "The forces contesting for the earth are so strong, so nearly matched, the battle so intense, that she could break apart under the strain. Feed the earth."

If you have come this far in this book, and done even a few of the practices suggested, you will have begun to experience the earth as a living being who is aware and speaking to us all the time. You will have glimpsed the miracle of her immense and intricate cycles, the tides and currents of atmosphere and ocean, her breath and blood, the passing of energy and nutrients through cycles of birth and death, decay and regeneration, the patterns that give form and substance to her flows and that form her body.

And if you have opened your awareness, you also know that she is hurting. The great cycles of the elements, the back-and-forth breath of green and red that infuses the atmosphere with oxygen, the diversity that has taken billions of years to evolve, are everywhere under assault. And we know that change may not be a gradual, slowly flowing river, but rather a stream that suddenly plunges down a rocky waterfall. At the moment, we can see and feel the earth's degradation, the unusual heat of this summer, the high rates of cancer, the loss of species—but her basic life-support systems are still functioning for us, and it's easy to close off awareness of the damage. We're like boaters drifting on a slow stretch of the stream, aware that we are descending but telling ourselves it's not so bad. Everything is normal, calm; there's nothing to get too worried about. Yet just ahead is the waterfall. We may not see it until we reach it, too late to turn back, with only time enough to cry out, "Why didn't we change course earlier?" as we plunge over the rapids.

The earth has great powers of resilience, but she is also fragile. The cycles we've learned about, the life-support systems, can adjust to immense impacts, but if they crash, they can start a cascading collapse that could have consequences devastating to much of the earth's life.

At the same time, great forces of love and healing also are growing in the world. We now have technologies that, should we actually use them, allow us to live lightly on the earth with comfort and security. We have knowledge and wisdom if we choose to apply them, about how to provide for human needs in ways that respect and enhance the balance of life. And we have a growing, global community of people committed to balanced ways of living.

In this crucial time, we are called to be healers—of the earth, of the human community, of each other and ourselves. We speak of "healing the earth," but in reality, what needs healing is our human relationship to the earth.

Healing begins with listening. Many years ago, in what now seems like a former lifetime, I was trained as a psychotherapist. Therapy/counseling is a discipline of listening. We open ourselves to our client's pains, fears, hopes, and experiences, and by listening and witnessing, we establish a healing relationship.

Earth-healing is a similar process. By following the practices of observation recommended here, by maintaining a personal practice of listening to the earth, we create a new relationship that is healing, not just for the earth, but for ourselves. For when we are out of communication with the elements and energies and processes that sustain our lives, we cannot be healthy or whole.

Listening is also a practical discipline. If I have a piece of land, a sick tree, a community to heal, I begin by listening in a particular way the Reclaiming community calls dropped and open attention.[2]

DROPPED AND OPEN ATTENTION: A HEALING MEDITATION

In the place that needs healing, ground and come into your senses. You might want to cast a simple circle of protection.

Take some time to observe this place, to notice the signs of ill health or of need, to also notice and give gratitude for the life, beauty, and diversity that exist.

Now sit comfortably or lie down, and close your eyes.

Imagine all your thoughts as a cloud of massed string around your head, and slowly begin winding them into a ball at the center of your head. Take your time, and when you are ready, notice how it feels to have your awareness focused and concentrated and centered in your head. Let the ball compress down to a point of light.

Now, breathing slowly, let that point of light, of awareness, begin to slowly drop down your spine. It floats down, breath by breath, coming to rest for a moment in your heart. Breathe deep and notice how it feels to have your awareness centered in your heart.

Now continue to breathe and let that point of light float down, down, on each breath. At last it comes to rest in your belly, just two inches below your navel. Notice how it feels to have your awareness centered in your gut-level intuition.

And now, as you breathe, let that point of light begin to expand, into a disk of light, an open awareness. Let it expand until it encompasses the edges of your physical body, and notice how it feels to let your physical reality come into your awareness.

And now let that plane of awareness expand until it encompasses the edges of your energy body, your aura, and notice how it feels to let your energetic body come into your awareness.

And now let it expand until it encompasses this place you sit in that needs healing. Take a moment, breathe deep, and notice how you feel.

Ask for information to come to you about what is needed here for healing. Stay open and breathing, and allow images and words and messages to form. You might meet someone here who can share wisdom or tell you what is needed.

After a time of openness, consider some of the following questions:

Is something lacking here?
Is there too much of something?
Is there something attacking or assaulting this system?
Is there something that needs to be released or gotten rid of?
Is there something that needs to be brought in?
Is there a healing image, sound, word, or symbol that can help strengthen the forces of life and vitality here?
Where can we find the resources needed for healing?
Are there helpers we can ask to support this healing work?

Sit with your answers, and when you are ready, thank any beings or sources of information you have encountered, and begin to draw your awareness in.

As you draw your awareness into your own energy body, consciously let go of any energies that are not yours. Imagine your aura as a filter that can exclude any toxic energies and anything that does not belong to you.

As you draw your awareness back to your physical body, again consciously let go of any toxic energies or potential physical manifestations of energies you have encountered. Consciously draw in energies of health and life.

Breathing deep, draw your awareness back into your belly until it becomes again a point of light. Now take a deep breath and let it go to wherever you want your awareness centered at this moment.

Visions of Health

My land in the coastal mountains is beautiful. Compared to the degraded landscapes most city dwellers are used to, it seems wild and natural and vibrant with energy. There are towering redwoods in the streambeds, resprouted from the stumps of giants logged decades ago. There are rolling hills that in the winter rains turn emerald green.

But if I travel a few miles down the road to the state park that preserves some stands of old-growth redwood, I can see that my land is impoverished, in a state of recovery after a century of abuse. The old trees dwarf my young redwoods. Their undergrowth is richer, denser. The streams are flowing with water and graced by a riparian vegetation that is varied and complex.

Without the old growth, I can have no picture in my mind of what true health would look like for this land. If we are to become earth-healers—creators of cultures of beauty, balance, and delight, as my friend Donald Engstrom says—we need some glimpses of what that state of health might look like.

My friend Penny Livingston-Stark and I teach courses in permaculture design and earth-healing together. Her acre of land in Point Reyes, California, serves for me as a vision of what health can be.

Penny and her partner, James, have transformed their modest ranch house. The interior is now covered in natural plasters, made of clay, sand, pigment, and flour paste, that glow in rich jewel tones. Just outside, lettuce and salad greens are growing near the door, along with herbs for cooking. Graywater from the house runs out into a biofilter of gravel planted with papyrus and cattails, and from there into a small, constructed stream that empties into a duckpond.

The ducks help keep down bugs and slugs in the garden. An arbor of natural poles supports grapes and provides shade beside the pond. The earth that was dug from the pond was used to build a cob and straw-bale office, a small building of undulating forms that looks over the pond. (Cob is a mixture of clay, straw, and sand that can be shaped into organic, flowing forms and can also be built into the strong walls of long-lasting structures.) The office walls curve into a sitting bench, where a clay oven in the form of a dragon can be fired up to bake pizza, bread, or salmon.

Further down in the garden, a fence of espaliered apples produces many varieties of fruit through the fall and summer. Chickens weed and fertilize the garden beds and provide eggs. An open-air, thatched bedstead made of Balinese bamboo invites visitors to lie down and rest. Two other small structures made of straw-bale and bamboo and covered with natural plasters serve as guest quarters.

Tucked around the edges of the garden are tomato beds, more fruit trees, potatoes growing out of brush piles, more ponds full of water plants and fish, and many sensual surprises. To visit her garden is truly to stroll through paradise, with fruit, flowers, food, and inspiration abounding everywhere.

City Repair

Last spring, my friend Delight and I drove up the Oregon Coast after priestessing a healing ritual for forest defenders in northern California who had been facing horrific violence from logging companies. Before our drive we had watched videos of young activists being beaten up in the trees; they were hog-tied and dangled by ropes 180 feet up in the air, and literally tortured in an attempt to remove them from the platforms where they sat to deter the cutting of some of the last remaining old-growth redwoods in the area. We drove along a route I had bicycled twenty years before, appalled now at seeing clearcut after clearcut, barren hillsides and denuded streambeds.

We reached Portland the next morning, heartsick and weary, and drove toward where we would be staying. As we pulled up, we found ourselves stopped at an intersection beside a lush and flourishing garden that seemed to have spilled out onto all the edges of the pavement. The center of the intersection was painted in a huge, colorful mandala that filled the street, its painted radial lines marking the four directions and drawing in energy. On one street corner, a group of people were happily working together to sculpt a bench with angel wings out of cob.

The bench stood beside a small, covered kiosk where a thermos of hot water stood for making tea, with mugs and teabags provided for anyone in the neighborhood who might feel thirsty. This echoes back to the revolutionary tradition of the Boston Tea Party, and is a living example of a Goddess/service gift economy.

On another street corner of this intersection, a small hut of branches that wrapped around a living tree made a playhouse for the neighborhood children. On a third, a sculpted stand made a beautiful giveaway box for neighbors to leave things they no longer needed or books they'd already read, alongside a bulletin board for posting notices of neighborhood events. On the fourth corner, a second cob bench was under construction.

When we stopped to take a closer look, we were warmly greeted by Janelle Kapoor, who was one of the many leaders overseeing construction at sites around the city as part of the Natural Building Convergence we had come to join. She took us over to a neighboring house, where a bread oven sculpted of cob was under construction at the edge of the driveway. The woman who owns the house is a healer, and she told us the oven was pointed to face the Goddess statue in her front yard, and was oriented to the street so that all the neighbors could use it.

At that point, I sat down and burst into tears. I was overwhelmed at the generosity of spirit I could feel embodied in all these projects, and that generosity was such a contrast to the greed and violence we had witnessed in the forest. This intersection in the outskirts of Portland had become a village center, a meeting ground for a community, and by its physical existence had called that sense of community into being.

The intersection was one of the first projects of a group called City Repair, whose goal it is to repair the broken links of community that underlie our alienated cities. Mark Lakeman, one of the group's founders, is a true magician—someone able to manifest ideas and principles in physical reality. He described to me their efforts.

> The whole transformed intersection is a self-service piazza, requiring no monetary economy to run. It is grown from the gift-economy model of the universe, and so will expand and reverse the conventional paradigm directly, and through stories that inspire, and also through replication, because people intuitively and consciously recognize it as habitat in the midst of confusion.

The grid pattern of our cities, Mark explained, is really a military pattern/colonial device, designed to allow maximum control of space and easy movement through it. But it is not designed for human interactions. "The grid also makes the earth measurable as a commodity unit," Mark said. "The grid regiments space in order to homogenize experience and imagination. Indeed, it is designed to discourage and eliminate interaction, so that 'Love your neighbor' remains a high ideal rather than a daily reality."

A village, in contrast, always has at least one central meeting ground, a place where people gather informally to share news and gossip, drink tea (or something stronger), fall in love, gather to make decisions, settle a quarrel, and so on.

"Remember that freedom of assembly requires a place to assemble," Mark continued. "Villages also have many outer, supporting places which invite and encourage social interaction. . . . Such environments never need to have anyone 'build community,' because the fabric supports interaction as a matter of the flow of life."

City Repair decided to create such meeting grounds in the city. Their first project was a teahouse-treehouse, shaped in the form of an enormous, translucent womb and built around a huge conifer. (The local indigenous trees of life are conifers.) The teahouse was constructed of Mark's lifelong collection of doors and windows, which he had amassed, as builders do, knowing that someday he would use them for something.

City Repair began holding gatherings in the teahouse, free of money as a means of exchange, on Monday evenings. Monday—Moon-day—was chosen in honor of the Goddess, to reclaim her day and make it sacred. The gatherings grew and grew until finally the city got word of this unpermitted structure and ordered it dismantled.

"It was a trap," Mark said with glee. "We lured them in to attack us, which charged up the 'hood. . . . And then the 'hood was ready to do more."

The group decided to escalate by taking over an intersection, holding a gathering, closing the streets, and beginning construction of a new project. The teahouse (locally called the T-Hows) transformed into a teahorse (or T-horse). They took a used pickup, built collapsible wings of bamboo and recycled plastic that unfolded into the fans/canopies of a gathering shelter, and began taking this teahorse to public spaces to serve as a mobile meeting ground.

"We took the teahorse around and around the whole city, in a huge circular pattern," he explained, "raising the issue of the absence of crossroads places, cultural nodes, meeting places, and colonialism in general. This led to an internal movement in Portland," which is now inspiring imitators in other places.

At first, of course, the city objected to the intersection repair, but over time some magic happened, and more enlightened members of the city government recognized that the project actually solved some of the problems their own urban planners had been struggling with: improving livability, making streets safer, building local culture, increasing communication, and slowing traffic in residential areas. Intersection repair, the reclaiming and restoring of the crossroads, became legalized, and new projects were started. They require approval from all four homeowners on the corners and from eighty percent of the surrounding neighbors.

Mark began his career as a corporate architect, then became disillusioned and went traveling in Europe, Egypt, New Zealand, and throughout Mexico and Central America in search of new visions to bring back for his community. In the Lacandon rainforest of the southern Mexican region of Chiapas, he met indigenous Mayan shamans who taught him that he needed to learn to listen to the earth and to his own deeper self, to stop talking and speak through actions.

> One of the things that was said to me when I was in the rainforest in southern Mexico is that saving the world isn't going to be about words, it's going to be about actions. As I was trying to recover, coming home and being in this deep culture shock, I was trying to understand what was meant by that. And I just started to initiate these projects. How can I work out what I want to say without words? It's got to be about manifesting through action, creating examples that give people a chance to inhabit a landscape that reflects what they're already thinking. Except you don't want to try to persuade them anymore, you don't want to try to convince or debate, you just want to create it and then people can understand, viscerally; then they can know better what they have been looking to create—to manifest now.
>
> When you are doing something for the community, never ask permission of external authorities. That's what one of my rainforest teachers taught me. You go to that edge where you need to get permission, and then you don't. It's that old saying about knowing yourself, who are you, what is it that is your right, what is your nature, your habitat. What is my habitat? That is the place that we are able to create from, out of our nature, and nobody can give permission to do that. You shouldn't ever ask for permission, and nobody can ever give permission to build your birthright. And that's not my wisdom speaking, that's indigenous knowledge 101.

City Repair now has projects in many different parts of Portland. One intersection is graced by a giant Fibonacci sunflower mandala with an echoing archway of welded metal suspended beside a mosaic fountain. Another holds a Chartres-style labyrinth.

As we stood in that intersection, the women from two of the corner houses came out to welcome us.

"I see myself as a guardian of the labyrinth now," one told me, "in the tradition of all the ancient priestesses."

"It's been a while since we've had guardians of labyrinths in our cities," I agreed, "but I hope this is the beginning of a trend."

"Each thing that we do is a foothold that is taking the consciousness of the people participating, and the people observing, and the people hearing all the stories higher and higher, farther and farther, returning in a sense to who we are," Mark said. "Spirituality to me is simply communication with other people in community."[3]

Healing our relationship to the place we live is the beginning of creating community, and a healthy community is the ground for healing the land and our relationship to the natural communities that surround us.

Cancun

I kept thinking of Mark's words, and the advice he'd received in the rain forest, when we were in Cancun in September of 2003 as part of a mobilization protesting the World Trade Organization. The shaman's advice was echoed in the words we heard from our Mexican student allies, who had been deeply influenced by the Zapatista movement, which itself originated in the Lacandon rain forests of Chiapas. The Zapatistas represent the indigenous peoples who rebelled against the imposition of corporate globalization on their ancient cultures and livelihoods. When the North American Free Trade Agreement was passed in 1994, they formed an army of rebellion. Since then, they have moved away from armed struggle and embraced a political process of change. They now control many of the forested areas of Chiapas, and their example and thinking have inspired the global justice movement.

"The Zapatistas tell us that our power is not just in confronting the police and military power of the state," Abram, one of the students from Mexico City, explained in a planning meeting. "Our power is in claiming our own autonomous spaces, and building the world that we want."

Magic also teaches us to channel and direct energy, and to do that we need to know what we *want*, not just what we *oppose*. Confrontation is often necessary and unavoidable if we want to preserve and protect the earth, but when we bring our creativity and vision into the points of conflict, transformative moments can occur.

In Cancun, the Green Bloc that I was a part of had come down with the intention of making our encampments as much as possible models of ecological design. We included students from our Earth Activist Trainings, permaculturalists from the U.S., and allies from Mexico. We have an ongoing relationship with Tierra Viva, a group of street youth and punks from the slums of Mexico City who are using permaculture to transform their neighborhoods and are teaching gardening to children. Rodrigo Castellano, from the Mexican ecology group Biosfera, came down to help as well, bringing a bag of compost worms with him through airport security. "They're not on the list of forbidden objects," he explained to the security guard. "They have no sharp edges."

The city of Cancun was providing camping space for the students and the *campesinos*, the farmworkers and indigenous people, who were coming to march and to protest the policies of the World Trade Organization, which open the world's resources to exploitation by corporations and undermine citizens' ability to regulate environmental, health, and labor standards. But the city had no money to provide amenities, beyond rows of portable toilets.

We persuaded them to let us create a model handwashing/dishwashing stand next to the food tents. We fit flexible pipes to the edges of the canopies to serve as gutters and catch rain, which was stored in a tank, then pumped up to an elevated container with a simple hand-pump that worked by running a rope fitted with pistons through pipes and around a bicycle wheel. The water then went into faucets and down through small "sinks" made of bright orange funnels, to be channeled into a model graywater system consisting of a series of half-barrels. The first was filled with wood chips and duff, to filter grease, and the others were filled with gravel and water plants, to provide habitat for bacteria to break down disease organisms and toxins. We also created a simple shower installation that ran from an elevated tank fitted for rainwater collection. The installation was decorated with colorful flags made by the art collective for the mobilization that were printed with jaguars and Mayan Gods, and we also mounted photographs and printed up informational material on all the systems in both Spanish and English. The whole installation had a cheerful, colorful air about it, like an elaborate toy, and attracted much attention.

As with any project, we faced many challenges. Our original visions far outstripped our time and resources. Almost as soon as we set up our rain catchment, the rain stopped. (We had city water as a backup: one of the principles of permaculture is redundancy—that is, always having more than one source to meet any crucial need.) We had many moments of exhaustion, discouragement, and near-panic along the way.

But our little project was amazingly empowering and successful, often in ways we had not anticipated. It provided a living example that Mexicans and "gringos" could work together in a way that was mutually inspiring and respectful. It gave the media something positive to write about and photograph in the lead-up to the demonstrations, something that clearly embodied the principles we were fighting for. The project also demonstrated how an integrated system can work and proved that it can be made of simple, low-cost materials. It provided models that could be taken back to communities and used. Abby, Juniper, Riverwind, Eileen, and Cole demonstrated that women can build and design and make things. And, not least, the project provided a place for people to wash their hands before meals and to take showers!

There were many wonderful moments at the handwashing stand. Almost as soon as we set up our simple pump, a woman came over and studied it intently. "We have no running water in our village," she told me, "and a pump costs four thousand dollars. But *this* would work!"

I will always treasure the sight of Emilio, one of our black-clad, pierced, spiked, and studded punks, explaining the system to a group of thirty *campesinos*. Tierra Viva received invitations from many communities to come down and lead workshops in permaculture. And all the effort and worry and sweat were justified when Erik described what happened one morning as he was washing his hands: a five-year-old campesina girl came up to him and began explaining the entire system, from the rain catchment to the graywater, in complete and accurate detail. And she clearly understood it all: how the system turned a potential problem—runoff rainwater from the canopy—into a resource—clean water for handwashing and dishwashing—and how that water was then conserved and reused to grow plants and to be infiltrated into the ground (where, had this been a permanent installation, it could have been used to nurture a garden). We had created a living example of real abundance, and done it in a way that embodied cooperation and community.

There were many confrontations during the week of actions—at the fence the police erected to keep demonstrators out of the hotel zone, at the security barriers near the conference center where the WTO met. A Korean farmer/organizer, Hyung Hai Lee, stabbed himself to death on the first day of actions as an ultimate act of protest, bringing home to us all that these issues are matters of life and death for farmers and workers around the world. He was influenced by the suicide of a close friend who had lost his land because of economic policies that make it hard for small farmers to make a living.

Farmer suicide is a worldwide epidemic: Vandana Shiva spoke at the teach-in organized during the week of Cancun actions by the International Forum on Globalization and mentioned one area of India where 650 farmers had committed suicide in one month. In the face of those grim statistics of death, we were glad to be able to create something to embody hope.

In the middle of the week, we were able to stage a nonviolent blockade under the walls of the conference center itself. We spent the day dressed as tourists, filtering through the security system in ones or twos, and then converged just as the delegates were coming back from dinner. A small group of Mexican students and internationals moved out onto the road to blockade it by simply sitting down. Behind them, a group of us moved in and began a spiral dance, singing in Spanish,

Somos el viento que sopla,
Al imperio que colapsa.

And in English,

We are the rising of the moon,
We are the shifting of the ground,
We are the seed that takes root
When we bring the fortress down.

As we danced, members of the Green Bloc appeared carrying two trees and a bag of seeds. ("How did you get those trees past security?" I asked Rodrigo later. "We just carried them in. When they asked, 'What are you doing,' I just said, 'We're carrying these trees,' and walked on. People don't expect you to be carrying trees." Clearly, the man has a gift for this sort of thing!)

We danced around the trees, as Erik and John Henry spilled seeds on the ground in a spiral *ofrenda*, an altar/offering in the center of our circle, and raised power to charge our vision of a living world.

The ministerial ultimately collapsed in disagreements between the powerful countries of the north and the developing countries of the global south, who walked out under the leadership of Kenya. Delegates from the south told our friends inside that the many demonstrations inside the conference, outside the walls, and in the streets had given them the support they needed to take a strong stand and resist bullying from the U.S. Our chant was prophetic: we had indeed become the fresh wind that can blow away systems of destruction and open space for new seeds to grow.[4]

Community Vision

The struggle to reclaim and heal the earth is going on in every community. In one sense, anything we do to strengthen our communities and create networks that allow sharing, support, and connection is an act of earth-healing. But we are also often faced with stopping exploitation and degradation of our home environments.

In the Cazadero Hills, where I live, we have fought many battles against thoughtless development and the despoiling of the wild nature we love. We have formed land-use councils, advocacy groups, and a community land trust. When we learned that our county was undergoing a revision of its general plan, which it does every thirty years, we wanted to participate in the process. But the public meetings were held in Santa Rosa, an hour and a half from our homes. They were often rescheduled or canceled abruptly, and at best they allowed for very limited public comment.

Increasingly frustrated, we decided to conduct our own planning process. In the spring of 2002, we held a community vision day, inviting all the people who lived in our area to come together and discuss what they wanted. We facilitated *creative* processes, asking people to draw and write their visions, and also *cognitive* processes, organizing people into groups that considered the areas under discussion in the general plan. Using the results from that day, we began to write up a report, and we circulated a petition asking for a limit on vineyard development (which has been encroaching on the wildlands in our area). In the fall, when the hundred-page report was ready, we hosted a Wild Foods Breakfast, inviting county planners, elected supervisors, and the press, to present our report with a bit of fanfare that we hoped would get it noticed.

Our process inspired many other groups in the county that work on land issues, and it got good coverage in all the small local papers and even some local TV. (We were unable to persuade the major newspaper, the *Santa Rosa Press Democrat*, to cover it, however. Communities attempting to control their own destiny and resources are not "news." We finally got a reporter to come by telling him we would present the report in the nude—however, he left quickly when we appeared fully clothed.)

Our community action was not as dramatic as the frontline actions in Cancun, but it, too, was powerful and important. Articulating our community vision was empowering for everyone who took part in it. Writing it up and putting it out as a document made it part of the public record and gave it a new status in the dialogue. We had responded to an undemocratic, exclusionary process by creating our own autonomous response, and in so doing had carved out a new political space. We are not indigenous to this land, but we like to think that we are becoming indigenous, by loving the land, learning its plants and trees and animals and needs, taking responsibility for its health, and safeguarding its future.

Feeding the Earth

There are many ways to give energy and nurturing back to the earth. When we raise energy in ritual, we can consciously return it to earth with the intention of strengthening her life force and resilience. When we compost our food wastes, build soil in our gardens, heal the erosion surrounding a stream, or tend a small patch of our own ground, we are helping her overall healing.

We, too, are part of the earth, so healing ourselves and our communities is part of healing the earth. But spending time in nature, developing our relationship with her, and developing our own practices of observation and gratitude will also help heal our own wounds and imbalances.

Bringing alive visions of health and balance, carrying vision and creativity even into those battles where we stand up against the destruction of the earth, can help us to sustain our energy and can empower us with a knowledge of what we are fighting *for*, not just *against*.

Hope and Courage

When I was nearly finished with this book, I took a weekend off to join with Pagans and forest defenders in the Cascadia bioregion of Oregon, to share magic and healing. We camped near a beautiful grove of old-growth cedars and Douglas firs, and I was able to spend a long time lying among the roots of the trees, just looking and listening.

"The forces of greed are strong," the forest said to me. "But don't forget that you have immense forces working with you for the healing of the earth."

At times the process of destruction seems so advanced that we may find it hard not to sink into despair. "Action is the antidote to despair" is an Earth First! slogan, but nonstop action can be draining and disempowering if it is not thoughtful, strategic, and effective. And yet we do have great forces working with us, those same creative powers that arranged the chlorophyll mandala and learned to use sunlight to make food, that have traded genes and information for billions of years, that grew the redwood and the cedar from instructions encoded in the microscopic double-spiral DNA crystal, that move the great currents of the air and the tides of the ocean.

The earth is alive, and I don't believe that she is suicidal. She took a great gamble with this big-brain experiment, but I do believe that the consciousness we need to temper our destructive potential is arising, now, and is ultimately far, far stronger than greed. Moreover, I believe that the earth wants us to play the role we have evolved to play, a role as important as that of any worm or soil bacterium: to be her consciousness, her mirror, her great admirer and appreciator, to cheer her on and to use our specifically human abilities to help restore and sustain her balance.

When we are working in service of the earth, we can ask those powers to be with us. Indeed, they want to be asked, need to be asked. So here is a prayer for help. You may use it as written or let it inspire your own words:

PRAYER FOR HELP IN EARTH-HEALING

Great forces of creativity, growth, and love, great inventive imagination that has grown the diversity of life, mystery of the unfolding of form from energy, great powers of the trees and the grasses, the sunlight and the rain, great currents that move the continents and tides of the ocean, I send you love and gratitude and ask for your help. I am about to do ________________. It seems impossible. It seems beyond my human strength. Please lend me some of your power. Open my eyes and ears to inspiration. Release the obstacles that confront me, and draw in the resources, the luck, the energies that I need. Help me to succeed beyond my expectations, for the healing of the earth. Blessed be.

The forces of love and greed are indeed contesting with one another, and both are immensely strong—so strong that almost anything could tip the balance. Everything we do right now is vitally important. Each act of healing is a weight on the side of life. The drama moves toward its climax, and any one of us could be the small stone that starts the avalanche.

What a great time to be alive!

BLESSING FOR EARTH-HEALERS

We give thanks for all those who are moved, in their lives, to heal and protect the earth, in small ways and in large. Blessings on the composters, the gardeners, the breeders of worms and mushrooms, the soil-builders, those who cleanse the waters and purify the air, all those who clean up the messes others have made. Blessings on those who defend trees and who plant trees, who guard the forests and who renew the forests. Blessings on those who learn to heal the grasslands and renew the streams, on those who prevent erosion, who restore the salmon and the fisheries, who guard the healing herbs and who know the lore of the wild plants. Blessings on those who heal the cities and bring them alive again with excitement and creativity and love. Gratitude and blessings to all who stand against greed, who risk themselves, to those who have bled and been wounded, and to those who have given their lives in service of the earth.

May all the healers of the earth find their own healing. May they be fueled by passionate love for the earth. May they know their fear but not be stopped by fear. May they feel their anger and yet not be ruled by rage. May they honor their grief but not be paralyzed by sorrow. May they transform fear, rage, and grief into compassion and the inspiration to act in service of what they love. May they find the help, the resources, the courage, the luck, the strength, the love, the health, the joy that they need to do the work. May they be in the right place, at the right time, in the right way. May they bring alive a great awakening, open a listening ear to hear the earth's voice, transform imbalance to balance, hate and greed to love. Blessed be the healers of the earth.

Notes

CHAPTER ONE

1. Dion Fortune, an occultist and author of the nineteenth and early twentieth centuries, originated this definition. Although I've been quoting it for twenty years or more, I've never been able to track the exact reference down.

2. David Clarke with Andy Roberts, *Twilight of the Celtic Gods: An Exploration of Britain's Hidden Pagan Traditions* (London: Blandford, Cassell, 1996), pp. 22–24.

3. Starhawk, *The Spiral Dance: A Rebirth of the Ancient Religion of the Great Goddess* (San Francisco: HarperSanFrancisco, 1979, 1999), p. 103.

4. Kat Harrison, personal communication, 1994.

5. Starhawk, *Webs of Power: Notes from the Global Uprising* (Gabriola Island, B.C.: New Society Publishers, 2002), pp. 161–162.

6. Quoted by Bev Ortiz, "Contemporary California Indian Basketweavers and the Environment," in Thomas C. Blackburn and Kat Anderson, eds., *Before the Wilderness: Environmental Management by Native Californians* (Menlo Park, CA: Ballena Press, 1993), p. 199.

7. Allan Savory, *Holistic Management: A New Framework for Decision Making* (Washington, D.C., and Covelo, CA: Island Press, 1999), pp. 20–21.

8. Connie Barlow, *Green Space, Green Time: The Way of Science* (New York: Springer-Verlag, 1997), pp. 14–15.

CHAPTER TWO

1. Erik Ohlsen, who coteaches Earth Activist Trainings with me and Penny Livingston-Stark, led permaculture workshops at the Welcome Center and later blockaded with the Green Bloc who occupied the community garden.

2. Vandana Shiva, "The Green Revolution in the Punjab," *Ecologist* 21, no. 2 (March–April 1991). The article, extracted from *The Violence of the Green Revolution:*

Ecological Degradation and Political Conflict in Punjab (Dehra Dun, India: Vandana Shiva, 1989), can also be found at http://livingheritage.org/green-revolution.htm.

3. Kenny Ausubel, *Seeds of Change* (San Francisco: HarperSanFrancisco, 1994), p. 74.

4. Quoted by Bev Ortiz in Blackburn and Anderson, eds., *Before the Wilderness*, pp. 195–196.

5. Press release, "New Report Challenges Fundamentals of Genetic Engineering: Study Questions Safety of Genetically Engineered Foods," Center for the Biology of Natural Systems, http://cbns.qc.edu/harperspressrelease.pdf, Jan. 15, 2002.

6. Barry Commoner, "Unraveling the Secret of Life: DNA Self-Duplication, the Basic Precept of Biotechnology, Is Denied," *Gene Watch* (May–June 2003). See also http://www.criticalgenetics.org/gene_watch_article.htm.

7. John Robbins, "A Biological Apocalypse Averted," *Earth Island Journal* 16, no. 4 (Winter 2001–2002). See also http://www.mindfully.org/GE/GE3/Apocalypse-Averted-Robbins.htm. The study upon which this article was based is M. T. Holmes, E. R. Ingham, J. D. Doyle, and C. S. Hendricks, "Effects of *Klebsiella planticola* SDF20 on Soil Biota and Wheat Growth in Sandy Soil," *Applied Soil Ecology* 11 (1999): 67–78. See also Elaine Ingham, "Ecological Balance and Biological Integrity: Good Intentions and Engineering Organisms That Kill Wheat," www.soilfoodweb.com or http://www.organicconsumers.org/ge/klebsiella.cfm. This article is adapted from a presentation the author gave on July 18, 1998, at the First Grassroots Gathering on Biodevastation: Genetic Engineering, in St. Louis, Missouri.

8. Luisah Teish, *Jambalaya: The Natural Woman's Book of Personal Charms and Practical Rituals* (San Francisco: HarperSanFrancisco, 1985), p. 54.

9. Jeannette Armstrong, interviewed in Derrick Jensen, *Listening to the Land: Conversations About Nature, Culture, and Eros* (San Francisco: Sierra Club Books), p. 295.

10. Clarke, with Roberts, *Twilight of the Celtic Gods*, p. 38.

11. Marija Gimbutas's major works on the Goddesses of Old Europe are:

The Gods and Goddesses of Old Europe (London: Thames & Hudson; Berkeley: Univ. of California Press, 1974).

The Language of the Goddess (San Francisco: HarperSanFrancisco, 1989).

The Civilization of the Goddess: The World of Old Europe (San Francisco: HarperSanFrancisco, 1991).

The Living Goddesses, edited and supplemented by Miriam Dexter Robbins (Berkeley and Los Angeles: Univ. of California Press, 1999).

12. Starhawk, "The Dismembering of the World," in *Truth or Dare: Encounters with Power, Authority, and Mystery* (San Francisco: HarperSanFrancisco, 1987), pp. 32–70.

13. Quoted from an interview with Marija Gimbutas in the video *Signs Out of Time: The Story of Archaeologist Marija Gimbutas*, by Donna Read and Starhawk (Belili Productions, 2003). See also www.belili.org.

14. Starhawk, *Dreaming the Dark: Magic, Sex, and Politics* (Boston: Beacon Press, 1982).

15. Donald Rumsfeld at a Department of Defense news briefing, Feb. 12, 2003. For a poetic setting of this quote, see http://pages.zdnet.com/sartre65/wrack/id34.html.

CHAPTER THREE

1. Permaculturalist Patrick Whitefield defined permaculture as the "art of creating beneficial relationships" in a guest lecture at an Earth Activist Training taught by Penny Livingston-Stark and me at Ragman's Lane Farm, Gloucestershire, England, in August 2002. Patrick is the author of *How to Make a Forest Garden* (Little Clyden Lane, Clanfield, Hampshire: Permanent Publications, 2000) and *Permaculture in a Nutshell* (Little Clyden Lane, Clanfield, Hampshire: Permanent Publications, 2000).

2. For a study linking glyphosate exposure to lymphoma, see Lennart Hardell and Mikael Eriksson, "A Case Control Study pf Non-Hodgkin Lymphoma and Exposure to Pesticides," *Cancer* Vol. 85, no. 6 (March 15, 1999). See also http://www.gene.ch/genet/1999/Jun/msg00012.html.

3. Mary Daly, *Beyond God the Father* (Boston: Beacon Press, 1973). See the discussion on page 40.

4. Donella Meadows, "Places to Intervene in a System," *Whole Earth* 91 (Winter 1997): 78–84.

CHAPTER FOUR

1. A good resource for ecological awareness in all the major religions is the Religion and Ecology Web site: http://hollys7.tripod.com/religionandecology/index.html. A bibliography on Judaism and ecology can be found at http://www.coejl.org/learn/bib_basic.shtml. See also Richard C. Foltz, Frederick M. Denny, and Azizan Baharuddin, eds., *Islam and Ecology: Bestowed Trust* (Cambridge, MA: Harvard Univ. Press, 2003).

2. For a complete listing of Matthew Fox's books, see: http://www.matthewfoxfcs.org/sys-tmpl/tipstricks/. A few of his relevant works are:

The Coming of the Cosmic Christ: The Healing of Mother Earth and the Birth of a Global Renaissance (New York: Harper & Row, 1980).

Creativity: Where the Divine and the Human Meet (New York: Tarcher, 2002).

Creation Spirituality: Liberating Gifts for the Peoples of the Earth (San Francisco: HarperSanFrancisco, 1991).

Original Blessing: A Primer in Creation Spirituality (New York: Tarcher, 2002).

Passion for Creation: The Earth-Honoring Spirituality of Meister Eckhart (Burlington, VT: Inner Traditions, 1980, 2000).

Wrestling with the Prophets: Essays on Creation Spirituality and Everyday Life (New York: Tarcher, 2000).

See also *Arthur Waskow, Seasons of Our Joy: A Handbook of Jewish Festivals* (Boston: Beacon Press, 1982).

3. James Lovelock, *The Ages of Gaia: A Biography of Our Living Earth* (New York: Bantam, 1990).

4. Elisabet Sahtouris, *Earthdance: Living Systems in Evolution* (Alameda, CA: Metalog Books, 1996); and *Gaia: The Human Journey from Chaos to Cosmos* (New York:

Pocket Books, 1989). See also Timothy Ferris, *The Whole Shebang* (New York: Touchstone, 1997).

I thank Brian Swimme for the many inspirational stories I've heard him tell in many years of association with Matthew Fox's Institute for Culture and Creation Spirituality.

5. It was Lynn Margulis's groundbreaking insight that *eukaryotes* evolved as symbiotic communities of simpler life-forms.

6. For his reading of and helpful comments on this chapter, I thank David Seaborg, evolutionary biologist, environmental activist, and founder and president of the World Rainforest Fund, dedicated to saving the rain forest.

CHAPTER FIVE

1. Jane Jacobs, *The Death and Life of Great American Cities* (New York: Vintage, 1992), pp. 50–51.

2. Starhawk, *Walking to Mercury* (New York: Bantam, 1994), pp. 304–305.

3. See also the discussion of anchoring in Starhawk and Hilary Valentine, *The Twelve Wild Swans: A Journey to the Realm of Magic, Healing, and Action* (San Francisco: HarperSanFrancisco, 2000), pp. 41–43.

CHAPTER SIX

1. Starhawk, *The Spiral Dance*, pp. 80–83.

2. Starhawk, *The Spiral Dance*, pp. 87–101.

3. Starhawk, *The Spiral Dance*, pp. 87–101; Starhawk and Valentine, *Twelve Wild Swans*, pp. 80, 167–171, and 240–241.

4. Starhawk and Valentine, *Twelve Wild Swans*, pp. 167–171.

CHAPTER SEVEN

1. Teaching with activist and deep ecologist John Seed and his partner, Ruth Rosenhak, recently in Australia, I discovered that they use their own version of this same story and a very similar meditation—a beautiful example of parallel evolution at work! If Gaia is whispering the same thing into the ears of two sources on opposite sides of the world, she must really want us to do this one.

2. David Abram, *The Spell of the Sensuous: Perception and Language in a More-Than-Human-World* (New York: Pantheon Books, 1996), p. 4.

3. Cities for Climate Protection (at http://www.iclei.org/co2/) is a campaign of the International Council on Local Environmental Initiatives: http://www.iclei.org/. It offers a framework and assistance for local governments to reduce greenhouse gas emissions and achieve sustainability.

CHAPTER EIGHT

1. Abram, *Spell of the Sensuous*, p. 7.

2. Charles Dickens, *David Copperfield* (New York: The Paddington Corporation, 1965), chap. 12, p. 185. First published 1849–1850.

3. Rainforest Action Network's U'Wa campaign home page is http://www.ran.org/ran_campaigns/beyond_oil/oxy/.

4. Abram, *Spell of the Sensuous*, p. 21.

5. Quoted by Bev Ortiz in Blackburn and Anderson, eds., *Before the Wilderness*, p. 196.

6. Allan Savory, *Holistic Management.*

7. Jeanette Armstrong, "Keepers of the Earth," in Theodore Roszak, Mary E. Gomes, and Allen D. Kanner, eds., *Ecopsychology: Restoring the Earth, Healing the Mind* (San Francisco: Sierra Club Books, 1995), p. 323.

8. To find a CSA near you, check the Community Supported Agriculture farms database at http://www.nal.usda.gov/afsic/csa/csastate.htm.

9. Henry T. Lewis, "Patterns of Indian Burning in California: Ecology and Ethnohistory," in Blackburn and Anderson, eds., *Before the Wilderness*, pp. 55–116.

10. Starhawk and Valentine, *Twelve Wild Swans*, 301–302.

11. Starhawk and Valentine, *Twelve Wild Swans;* see "Anger Ritual," p. 102, and "Rage Ritual," p. 138.

CHAPTER NINE

1. Note on the chants used here: "Born of water" was created in a workshop I did with Kate Kaufman sometime in the early 1980s in Madison, Wisconsin. "I am the laughing one" and "I am the shaper" are variations that someone began singing in the middle of the trance at the British Columbia Witch Camp sometime in the late 1980s. I'm sorry I don't know the names of their creators. "The river is flowing" is by Adele Getty. "The ocean is the beginning of the earth" was made up by Delaney Johnson and me when he was six years old and I was thirty-five, and we sat overlooking the water on the Mendocino headlands.

2. Paul Simon, *Tapped Out: The Coming World Crisis in Water and What We Can Do About It* (New York: Welcome Rain, 1998), p. 21.

3. Marq de Villiers, *Water: The Fate of Our Most Precious Resource* (Boston and New York: Houghton Mifflin, 2001), p. 44.

4. Villiers, *Water,* p. 37.

5. Maude Barlowe, "The Global Water Crisis and the Commodification of the World's Water Supply," Spring 2001, http://www.canadians.org/display_document.htm?COC_token=COC_token&id=245&isdoc=1&catid=78.

6. Toby Hemenway, *Gaia's Garden: A Guide to Home-Scale Permaculture* (White River Junction, VT: Chelsea Green Publishing, 2001), pp. 89–90.

7. Alice Outwater, *Water: A Natural History* (New York: Basic Books, 1996).

8. See the Web site for Ocean Arks International at http://www.oceanarks.org/.

9. See Penny Livingston-Stark's Web site at http://www.permacultureinstitute.com.

10. Barlowe, "The Global Water Crisis."

11. The Cochabamba Declaration can be found at http://www.nadir.org/nadir/initiativ/agp/free/imf/bolivia/cochabamba.htm.

CHAPTER TEN

1. Gretel Erlich, quoted in Robert Clark, ed., *Our Sustainable Table* (San Francisco: Northpoint Press, 1990); see also http://www.public.iastate.edu/~vwindsor/Cross.html.

2. See Oregon Department of Agriculture statistics at http://www.oda.state.or.us/information/news/Food_spending.html.

3. Elaine Ingham, "Ecological Balance and Biological Integrity," www.soilfoodweb.com.

4. See, for example, the Web site of the Biodynamic Farming and Gardening Association: http://www.biodynamics.com/index.html.

5. Paul Stamets's Web site is http://www.fungiperfecti.org.

6. Kenny Ausubel, *Seeds of Change*, p. 64.

7. For more information about this, see Danielle Goldberg, "Jack and the Enola Bean," TED (Trade Environment Database) Case Study Number XXX 2003; www.american.edu/TED/enola-bean.htm.

8. "Free Tree, Free Tree," *Hindustan* (India) *Times*, June 9, 2000; at http://www1.hindustantimes.com/nonfram/090600/detOPI01.htm.

9. Press release, "New Report Challenges Fundamentals of Genetic Engineering: Study Questions Safety of Genetically Engineered Foods," Center for the Biology of Natural Systems, at http://cbns.qc.edu/harperspressrelease.pdf, Jan. 15, 2002.

10. See at http://www.percyschmeiser.com/.

CHAPTER ELEVEN

1. For information on the video and a complete bibliography of Gimbutas's work, see www.gimbutas.org/.

See the discussion of her work in J. Marler, ed., *From the Realm of the Ancestors: An Anthology in Honor of Marija Gimbutas* (Manchester, CT: Knowledge, Ideas, and Trends, 1997).

See also Carol Christ, *Rebirth of the Goddess: Finding Meaning in Feminist Spirituality* (Reading, MA: Addison-Wesley, 1997), pp. 70–88, and David Miller, *I Didn't Know God Made Honky Tonk Communists* (Berkeley, CA: Regent Press, 2002), pp. 141–155.

2. Tyler Volk, *Metapatterns: Across Space, Time, and Mind* (New York: Columbia Univ. Press, 1995), pp. 12–13.

3. Alice Outwater, *Water*, p. 57.

4. For directions on building an herb spiral, see Bill Mollison, *Introduction to Permaculture* (Tyalgum, Australia: Tagari, 1991), p. 96; and Toby Hemenway, *Gaia's Garden*, pp. 48–49.

5. Christopher Alexander, Sara Ishikawa, and Murray Silverstein, with Max Jacobson, Ingrid Fiksdahl-King, and Shlomo Angel, *A Pattern Language: Towns, Buildings, Construction* (New York: Oxford Univ. Press, 1977), pp. 618–621.

6. Tyler Volk, *Gaia's Body: Toward a Physiology of Earth* (New York: Copernicus, 1988), p. 128.

7. For examples and much more information on labyrinths, see Lauren Artress, *Walking a Sacred Path: Exploring the Labyrinth as a Spiritual Tool* (New York: Riverhead Books, 1996); and Sig Lonegren, *Labyrinths: Ancient Myths and Modern Uses* (Glastonbury: Gothic Image Publications, 1991, 1996).

8. Diane Baker, Anne Hill, and Starhawk, *Circle Round: Raising Children in Goddess Tradition* (New York: Bantam, 1998); and Starhawk, *The Spiral Dance*, pp. 193–213.

CHAPTER TWELVE

1. This chant was written by Adele Getty.

2. I credit Cybele for introducing me and the broader Reclaiming community to the practice of dropped and open attention, and she credits Wendy Palmer, *The Intuitive Body: Aikido as a Clairsentient Practice* (Berkeley, CA: North Atlantic Books, 2000).

3. Mark Lakeman's statements come from a personal interview with him I conducted in August of 2003. The City Repair Web site is www.cityrepair.org.

4. My daily updates from Cancun are posted on my Web site, www.starhawk.org. The chant was written by me and Rodrigo Castellano.

Select Bibliography

Works I've consulted in writing this book, and a small sampling of the many works that may be helpful to the reader wanting to explore these issues further.

David Abram. *The Spell of the Sensuous: Perception and Language in a More-Than-Human World*. New York: Pantheon Books, 1996.

Christopher Alexander, Sara Ishikawa, and Murray Silverstein, with Max Jacobson, Ingrid Fiksdahl-King, and Shlomo Angel. *A Pattern Language: Towns, Buildings, Construction*. New York: Oxford Univ. Press, 1977.

Luke Anderson. *Genetic Engineering, Food, and Our Environment*. White River Junction, VT: Chelsea Green Publishing, 1999.

Lauren Artress. *Walking a Sacred Path: Exploring the Labyrinth as a Spiritual Tool*. New York: Riverhead Books, 1996.

Kenny Ausubel. *Restoring the Earth: Visionary Solutions from the Bioneers*. Tiburon: H. J. Kramer, 1997.

Kenny Ausubel. *Seeds of Change*. San Francisco: HarperSanFrancisco, 1994.

Albert-Laszlo Barabasi. *Linked: The New Science of Networks*. Cambridge, MA: Perseus Publishing, 2002.

Michael Barbour, Bruce Pavlik, Frank Drysdale, and Susan Lindstrom. *California's Changing Landscapes: Diversity and Conservation of California Vegetation*. Sacramento: California Native Plant Society, 1993.

Connie Barlow. *Green Space, Green Time: The Way of Science*. New York: Springer-Verlag, 1997.

Maude Barlowe and Tony Clarke. *Blue Gold: The Battle Against Corporate Theft of the World's Water*. Toronto: Stoddart, 2002.

Maude Barlowe and Tony Clarke. *Global Showdown*. Toronto: Stoddart, 2001.

John J. Berger, ed. *Environmental Restoration: Science and Strategies for Restoring the Earth*. Washington, D.C.: Island Press, 1990.

Thomas C. Blackburn and Kat Anderson, eds. *Before the Wilderness: Environmental Management by Native Californians*. Menlo Park, CA: Ballena Press, 1993.

Carol Christ. *Rebirth of the Goddess: Finding Meaning in Feminist Spirituality*. Reading, MA: Addison Wesley, 1997.

Diana Leafe Christian. *Creating a Life Together: Practical Tools to Grow Ecovillages and Intentional Communities*. Gabriola Island, B.C.: New Society Publishers, 2003.

David Clarke, with Andy Roberts. *Twilight of the Celtic Gods: An Exploration of Britain's Hidden Pagan Traditions*. London: Blandford, 1996.

Ronnie Cummins and Ben Lilliston. *Genetically Engineered Food: A Self-Defense Guide for Consumers*. New York: Marlowe & Co., 2000.

Mary Daly. *Beyond God the Father*. Boston: Beacon Press, 1973.

Richard Dawkins. *Climbing Mount Improbable*. New York: Norton, 1996.

Irene Diamond and Gloria Orenstein, eds. *Reweaving the World: The Emergence of Ecofeminism*. San Francisco: Sierra Club Books, 1990.

Charles Dickens, *David Copperfield*. New York: The Paddington Corporation, 1965; first published 1849–1850.

Alan Drengson and Duncan Taylor. *Ecoforestry: The Art and Science of Sustainable Forest Use*. Gabriola Island, B.C.: New Society Publishers, 1997.

David Duhon. *One Circle: How to Grow a Complete Diet in Less than 1000 Square Feet*. Willits, CA: Ecology Action, 1995.

Joan Dunning. *From the Redwood Forest: Ancient Trees and the Bottom Line—A Headwaters Journey*. White River Junction, VT: Chelsea Green Publishing, 1998.

Timothy Ferris. *The Whole Shebang*. New York: Touchstone, 1997.

Richard C. Foltz, Frederick M. Denny, and Azizan Baharuddin, eds. *Islam and Ecology: Bestowed Trust*. Cambridge, MA: Harvard Univ. Press, 2003.

Matthew Fox. *The Coming of the Cosmic Christ: The Healing of Mother Earth and the Birth of a Global Renaissance*. Scranton, PA: Harper & Row, 1980.

Matthew Fox. *Creation Spirituality: Liberating Gifts for the Peoples of the Earth*. San Francisco: HarperSanFrancisco, 1991.

Matthew Fox. *Creativity: Where the Divine and the Human Meet*. New York: Jeremy P. Tarcher, Inc. 2002.

Matthew Fox. *Original Blessing: A Primer in Creation Spirituality*. New York: Jeremy P. Tarcher, Inc. 2002.

Matthew Fox. *Passion for Creation: The Earth-Honoring Spirituality of Meister Eckhart*. Vermont: Inner Traditions, 1980, 2000.

Matthew Fox. *Wrestling with the Prophets: Essays on Creation Spirituality and Everyday*. New York: Jeremy P. Tarcher, 2000.

Marija Gimbutas. *The Civilization of the Goddess: The World of Old Europe*. San Francisco: HarperSanFrancisco, 1991.

Marija Gimbutas. *The Gods and Goddesses of Old Europe*. London: Thames & Hudson; Berkeley, CA: Univ. of California Press, 1974.

Marija Gimbutas. *The Language of the Goddess*. San Francisco: Harper & Row, 1989.

Marija Gimbutas (edited and supplemented by Miriam Dexter Robbins). *The Living Goddesses*. Berkeley and Los Angeles: Univ. of California Press, 1999.

Brian Greene. *The Elegant Universe: Superstrings, Hidden Dimensions, and the Quest for the Ultimate Theory*. New York: W.W. Norton, 1999.

Jesse Wolf Hardin. *Gaia Eros: Reconnecting to the Spirit of Nature*. Reserve, NM: Earthen Spirituality Project and Sweet Medicine Women's Center, 2003; at earthway@concentric.net.

Richard Heinberg. *The Party's Over: Oil, War, and the Fate of Industrial Society*. Gabriola Island, B.C.: New Society Publishers, 2003.

Toby Hemenway. *Gaia's Garden: A Guide to Home-Scale Permaculture*. White River Junction, VT: Chelsea Green Publishing, 2001

Erich Hoyt and Ted Schultz, eds. *Insect Lives: Stories of Mystery and Romance from a Hidden World*. New York: Wiley, 1999.

Jane Jacobs. *The Death and Life of Great American Cities*. New York: Vintage, 1992.

Jane Jacobs. *The Nature of Economies*. New York: Modern Library, 2000.

John Jeavons. *How to Grow More Vegetables*. Berkeley, CA: Ten Speed Press, 1979.

Steven Johnson. *Emergence: The Connected Lives of Ants, Brains, Cities, and Software*. New York: Simon & Schuster, 2001.

David Korten. *The Post-Corporate World: Life After Capitalism*. San Francisco: Berrett-Koehler; West Hartford, CT: Kumarian, 1999.

David Korten. *When Corporations Rule the World*. San Francisco: Berrett-Koehler; West Hartford, CT: Kumarian, 1995.

Robert Kourik. *Designing and Maintaining Your Edible Landscape Naturally*. Santa Rosa, CA: Metamorphic Press, 1986.

Aldo Leopold. *A Sand County Almanac*. New York: Oxford Univ. Press, 1966.

Sig Lonegren. *Labyrinths: Ancient Myths and Modern Uses*. Glastonbury: Gothic Image, 1991, 1996.

Joanna Macy. *Thinking Like a Mountain: Toward a Council of All Beings*. Philadelphia, PA: New Society Publishers, 1988.

Malcolm Margolin. *The Way We Lived: California Indian Reminiscences, Stories, and Songs*. Berkeley, CA: Heyday Books, 1981.

Alastair McIntosh, *Soil and Soul: People Versus Corporate Power*. London: Aurum Press, 2001.

David Miller. *I Didn't Know God Made Honky Tonk Communists*. Berkeley, CA: Regent Press, 2002.

Bill Mollison. *Introduction to Permaculture*. Tyalgum (Australia): Tagari, 1991.

Bill Mollison. *Permaculture: A Designer's Manual*. Tyalgum (Australia): Tagari, 1988, 1992.

Bill Mollison, with Reny Mia Slay. *Introduction to Permaculture*. Tyalgum (Australia): Tagari Press, 1995.

Bill Moyer. *Doing Democracy: The MAP Model for Organizing Social Movements*. Gabriola Island, BC: New Society Publishers, 2001.

Beverly R. Ortiz, as told by Julia F. Parker. *It Will Live Forever: Traditional Indian Acorn Preparation*. Berkeley, CA: Heyday Books, 1991.

Alice Outwater. *Water: A Natural History*. New York: Basic Books, 1996.

Wendy Palmer. *The Intuitive Body: Aikido as a Clairsentient Practice*. Berkeley, CA: North Atlantic Books, 2000.

Marc Reisner. *Cadillac Desert: The American West and Its Disappearing Water*. New York: Penguin, 1986.

Jeremy Rifkin. *The Hydrogen Economy*. New York: Jeremy P. Tarcher, 2002.

Carolyn Roberts. *House of Straw: A Natural Building Odyssey*. White River Junction, VT: Chelsea Green Publishing, 2002.

Theodore Roszak, Mary Gomes, and Allen D. Kanner, eds. *Ecopsychology: Restoring the Earth, Healing the Mind*. San Francisco: Sierra Club Books, 1995.

Elisabet Sahtouris. *Earthdance: Living Systems in Evolution*. Alameda, CA: Metalog Books, 1996.

Elisabet Sahtouris. *Gaia: The Human Journey from Chaos to Cosmos*. New York: Pocket Books, 1989.

Greg Sarris. *Mabel McKay: Weaving the Dream*. Berkeley and Los Angeles: Univ. of California Press, 1994.

Allan Savory. *Holistic Management: A New Framework for Decision Making*. Washington, D.C./Covelo, CA: Island Press, 1999.

Randy Shaw. *The Activist's Handbook*. Berkeley and Los Angeles: Univ. of California Press, 1996.

Rupert Sheldrake. *The Presence of the Past*. London: Collins, 1988.

Rupert Sheldrake. *The Rebirth of Nature: The Greening of Science and God*. New York: Bantam, 1991.

Alix Kates Shulman. *Drinking the Rain*. New York: Farrar, Straus, Giroux, 1995.

Michael G. Smith. *The Cobber's Companion: How to Build Your Own Earthen Home*. Cottage Grove, OR: The Cob Cottage, 1998.

Starhawk. *Dreaming the Dark: Magic, Sex, and Politics*. Boston: Beacon Press, 1982.

Starhawk. *The Fifth Sacred Thing*. New York: Bantam, 1992.

Starhawk. *The Spiral Dance: A Rebirth of the Ancient Religion of the Great Goddess*. San Francisco: HarperSanFrancisco, 1979, 1999.

Starhawk. *Truth or Dare: Encounters with Power, Authority, and Mystery*. San Francisco: HarperSanFrancisco, 1987.

Starhawk. *Walking to Mercury*. New York: Bantam, 1994.

Starhawk. *Webs of Power: Notes from the Global Uprising*. Gabriola Island, B.C.: New Society Publishers, 2002.

Starhawk, Diane Baker, and Anne Hill. *Circle Round: Raising Children in Goddess Tradition*. New York: Bantam, 1998.

Starhawk and M. Macha Nightmare. *The Pagan Book of Living and Dying*. San Francisco: HarperSanFrancisco, 1997.

Starhawk and Hilary Valentine. *The Twelve Wild Swans: A Journey to the Realm of Magic, Healing, and Action*. San Francisco: HarperSanFrancisco, 2000.

Sara Stein. *Noah's Garden: Restoring the Ecology of Our Own Back Yards*. Boston: Houghton Mifflin, 1993.

Sara Stein. *Planting Noah's Garden: Further Adventures in Backyard Ecology*. Boston: Houghton Mifflin, 1997.

Brian Swimme. *The Hidden Heart of the Cosmos*. Maryknoll, NY: Orbis Books, 1999.

Brian Swimme and Thomas Berry. *The Universe Story*. San Francisco: Harper-SanFrancisco, 1994.

Luisah Teish: *Jambalaya: The Natural Woman's Book of Personal Charms and Practical Rituals*. San Francisco: Harper & Row, 1985.

Nancy Jack Todd and John Todd. *From Eco-Cities to Living Machines: Principles of Ecological Design*. Berkeley, CA: North Atlantic Books, 1994.

Brian Tokar, ed. *The Worldwide Challenge to Genetic Engineering*. London: Zed Books, 2001.

Sim Van der Ruyn and Stuart Cowan. *Ecological Design*. Washington D.C.: Island Press, 1996.

Sim Van der Ruyn. *The Toilet Papers: Recycling Waste and Conserving Water*. Sausalito, CA: Ecological Design Press, 1978.

Tyler Volk. *Gaia's Body: Toward a Physiology of Earth*. New York: Copernicus, 1998.

Tyler Volk. *Metapatterns: Across Space, Time, and Mind*. New York: Columbia Univ. Press, 1995.

Alan Weisman. *Gaviotas: A Village to Reinvent the World*. White River Junction, VT: Chelsea Green Publishing, 1995.

Patrick Whitefield. *How to Make a Forest Garden*. Little Clyden Lane, Clanfield, Hampshire: Permanent Publications, 2000.

Patrick Whitefield. *Permaculture in a Nutshell*. Little Clyden Lane, Clanfield, Hampshire: Permanent Publications, 2000.

Edward O. Wilson. *The Diversity of Life*. New York: Norton, 1992.

Linda Woodrow. *The Permaculture Home Garden*. Ringwood, Victoria (Australia): Viking Penguin, 1996.

Resources

Starhawk's Projects

STARHAWK'S WEB SITE:
www.starhawk.org

ROOT ACTIVIST NETWORK OF TRAINERS
www.rantcollective.org
(Starhawk is a member of this collective, which offers trainings in organizing and nonviolent direct action.)

EARTH ACTIVIST TRAINING
21 Fort Ross Way
Cazadero, CA 95421
707-583-2300, ext. 119
www.earthactivisttraining.org/
(These trainings combine a permaculture design course with activist and organizing skills and earth-based spirituality.)

BELILI PRODUCTIONS
www.belili.org
(Starhawk and Donna Read produce documentaries on issues related to earth-based spirituality, women, and social change. Their first project is *Signs Out of Time: The Story of Archaeologist Marija Gimbutas*. They are currently working on a documentary about permaculture.)

Pagan Resources

RECLAIMING
P.O. Box 14404
San Francisco, CA 94114
www.reclaiming.org
(Reclaiming is the group Starhawk cofounded over twenty years ago. It offers training in Goddess-based and earth-based spirituality integrated with social change. Reclaiming groups in North America, Europe, and elsewhere offer a variety of resources, including public rituals, courses, Witch camps, and gatherings.)

THE WITCHES' VOICE
www.witchvox.com
(This is the most comprehensive of the hundreds of Pagan Web sites.)

Permaculture Resources

CITY REPAIR
1237 SE Stark
P.O. Box 42615
Portland, OR 97242
503-235-1046
thecircle@cityrepair.org
www.cityrepair.org

PERMACULTURE INSTITUTE OF NORTHERN CALIFORNIA
P.O. Box 341
Point Reyes Station, CA 94956
415-663-9090
www.permacultureinstitute.com
(Penny Livingston-Stark, Starhawk's teaching partner, is codirector of PINC.)

THE PERMACULTURE ACTIVIST
P.O. Box 1209
Black Mountain, NC 28711
voicemail: 828-669-6336; fax: 828-669-5068
www.permacultureactivist.net
(This is the central clearinghouse for permaculture information and contacts in North America, and publisher of the *Permaculture Activist* magazine.)

PERMACULTURE CREDIT UNION
4250 Cerrillos Road
P.O. Box 29300
Santa Fe, NM 87592-9300
505-954-3479; toll-free: 866-954-3479
fax: 505-424-1624
www.pcuonline.org/

Wilderness Awareness Resources

WILDERNESS AWARENESS SCHOOL
www.natureoutlet.com

JON YOUNG'S TRACKING SCHOOL
www.shikari.org

TOM BROWN'S TRACKING SCHOOL
www.trackerschool.com